D0521640

BY THE BOARD OF HEALTH.

BOSTON, *October 2, 1799*.

AT a meeting of the *Board of Health* this day, *Voted*—That as the feafon has now arrived when the emptying of the contents of privies will be the leaft offenfive to the inhabitants, it be earneftly recommended to and required of the inhabitants, immediately to empty all Privies whofe contents are within 18 inches of the furface.

Voted—That no Privy within the town fhall be emptied from and after the publication of this order, except by a perfon or perfons licenced therefor by this Board; and every inhabitant intending to open or cleanfe a Privy, fhall alfo apply to them for a licence therefor, to empty the fame under fuch reftrictions and regulations as they may from time to time direct; and

Notice is hereby given to the Inhabitants of Bofton, that fuitable carts for the purpofe of tranfporting the contents of Privies to the places affigned therefor, have been purchafed by this Board, at the expenfe of the town, and it is voted, that no others fhall be permitted to be ufed by the inhabitants for that purpofe; but fuch of them as wifh to ufe tight Hogfheads have liberty fo to do, provided they caufe them to be immediately carried out of the town, or thrown direct into the channel unopened.

By Order of the Board,

PAUL REVERE, Prefident.

Atteft—J. W. Folsom, Secretary.

AND OTHER VARMINTS

Richard L. Carlton

Yes, the proverbial garden path was rife with danger, a pitch black privy interior could quickly change a faint-hearted depositor's mind. Children harbored the unspoken fear of dropping through an adult-sized seat hole or of being attacked by undetected insects from below. Snakes lurked in weeds along the way, and occasionally crawled into the cool dark chamber to be with you. In short, *you were vulnerable,* especially if caught with your pants around your ankles.

Among other prospective guests one might involuntarily entertain were wasps, hornets, bees, bats, rats, lizards, scorpions and skunks; and for some strange unexplained reason, hungry porcupines who often made after hours visits to gnaw away the edges of seat openings.

One middle-aged matron wrote to us about her experience with an "albino" spider! A large, hairy tarantula had been knocked into the privy pit by a previous sitter... and vengefully covered with powdered white lime. This huge white arachnid rose from the dead the next morning and crawled up the wall to abruptly confront our sleepy-eyed informant, whose hysterical cries for help brought heavily-armed men running from every corner of the campground.

There is no doubt that indoor plumbing was indeed welcomed by all right-minded people. But the passing of time dulls unpleasant memories and enhances good ones. How boring is modern-day life, especially the elimination aspect! In "the good old days" a trip to the backhouse was an adventure, a communion with nature, fraught with danger and excitement. Forget the rat snake that had you standing on the privy seat for half an hour. Forgive the neighbor's prize bull who fell asleep one summer afternoon against the outhouse door, leaving you trapped inside for three hours! Blot out from memory that awful stench which made you spit three times along the path back to Grandma's dinner table. Omit the smarting hornet's sting and the black fly's pesky bite. Ponder instead on the innocent discoveries of youth... the shared gossip, Sunday socials, hayloft frolics, secret swimming holes, and lost loves of a simpler, if not happier time.

UNCLE DUM DUM & HIS TOOTHPICK

by Victor Scott

Uncle Dum Dum got his nickname from a nervous habit of always drumming on the table or arm of his chair and humming, "dum-dum-dummity-dum." His real name was William Rupert Scott.

Uncle Bill was a drummer, a traveling salesman they would call him now. Every few months he would pass through a neighboring village and stay over at our house for a day or two.

We were always happy to see Uncle Bill. He could make all kinds of toys out of wood — and one of the things he sold was candy. He never failed to have a few samples with him. He wore a gold chain across his vest. On one end of the chain was a gold watch, and on the other end, a solid gold toothpick! When he pushed his chair back from the table, we would watch with awe as he picked his teeth with this piece of pointed jewelery.

Dum Dum was always in a hurry; except when playing with us kids. When he went to our outhouse, he went in a hurry, and never stayed long; except the time he had traveled all night — he claimed a crying baby on the train had kept him awake. After finishing his dinner Uncle Bill took the gold toothpick from its chain and wandered out to the privy, picking his teeth, along the way. He stayed . . . and stayed . ., until we all became worried!

At last Dad had me go knock on the door and see what was wrong. Uncle Bill came out rubbing his eyes, he had fallen asleep. Lack of rest, droning bees, and the smell of apple blossoms had done their work.

Bleary eyed Uncle Bill got about halfway back to the house and yelled suddenly, "My toothpick!" He ran back and looked where he had been sitting — and sleeping. He searched frantically for a while, then, giving up, turned to Jimmy and me, saying, "Boys, I'll give you fifty cents each if you will find that toothpick!"

"Go ahead and search for it," Dad said, "I'll let you off from work this afternoon while you are looking." We walked over to the outhouse and near the door I immediately saw something glinting in the grass. I reached down and picked up the toothpick, and started running to the house; but Jimmy stopped me. "If we find it this soon, we'll have to go to the field and work all day." Reason won out over instinct!

We managed to fool a few hours away, and then proudly went to the house and gave our uncle the toothpick. He pulled out two fifty cent pieces from his pocket and handed them to us. Our eyes were bigger than the half dollars. Neither of us had ever owned one.

"Boys, let's go fishing", he said. It's too late to go to the field. I'll fix it with your dad." When we got back, Uncle Bill scrubbed and scrubbed the toothpick and put it back on the chain.

After supper he pushed back his chair and started to pick his teeth, but he hesitated for a moment and held the sullied probe up to the light. "Don't be afraid," I said, "we found it in the grass." He smiled and continued picking his teeth.

THIS SIDE OF OLD FAITHFUL

by Victor Scott

On long summer days when father and the older boys were working in the field, there would be hours when our outhouse was not in use in the regular sense of the word. Sometimes the little girls used it for a playhouse — with the understanding, of course, that they might have to grab their dolls and move at a moment's notice. The family privy also served as a fort during corncob fights — not to get inside, but to get behind and make short forays at the enemy.

A scaly bark hickory tree grew with its branches spreading over "Old Faithful". In the autumn on a windy day, the hickory nuts sounded like bullets hitting the tin roof of our fort as we sat there looking through the cracks which served as loopholes to shoot imaginary enemies.

The apple orchard was behind it, and even today the scent of apple blossoms or ripe apples brings back vivid memories. It seemed the natural thing to sit on the smaller of two holes and chew on green apples.

Two incidents stand out in memory concerning the outhouse. One is about the time we discovered Aunt Midian was "human". She was given that name one morning when three of us boys were practicing shooting marbles on the hard packed ground of the path leading to "Old Faithful". Suddenly Jimmy grabbed his marbles and jumped to the side of the path, hissing, "Midian is coming! Everyone flee to his tent, Oh Israel!"

If we had not jumped aside we probably would have been stepped on. In her usual grim manner, looking neither right, nor left, our visiting aunt was making her way, skirts billowing under full sail, for the outhouse.

The name "Midian" stuck. We were all a little afraid of her. She was a large woman, and seemed to have a perpetual grim look about her. When she came to visit, we were supposed to line up for her to kiss us. I often wondered if she hated the ordeal as much as we did. We talked among ourselves, wondering if she ever really *had* to use the outhouse, or just stayed in it for long periods of time to keep anyone else from using it. The suggestion was made, and partially believed by the smaller children, that perhaps she was not human!

One day late in autumn, Aunt Midian said she would be leaving early next morning. Later in the afternoon I was cracking hickory nuts with my two little sisters under the scaly bark tree and picking out kernels with horseshoe nails. I would crack nuts and help little Jean pick out kernels — she was only five. Mary Belle was eight and could do a pretty good job, but I had to recrack several for her.

Along came Aunt Midian, her skirts billowing in the breeze. Looking straight ahead she seemed not to see us sitting a few feet from the path. She went inside and slammed the door. Jean immediately decided she needed to go, but knowing from experience it would be a long time, I asked Mary Belle to take her out in the bushes. They were back in about a minute and Mary Belle said, "She really didn't need to go. It was just physiology because Aunt Midian was inside."

When Midian came out, she glanced at us but didn't say a word. As she went past we couldn't believe our eyes! Her dress was caught up in the biggest pair of pink bloomers I had ever seen in my life! They were the size Dad would describe as two ax handles wide.

"Oh my...she don't know!" Mary Belle exclaimed as she ran up to Aunt Midian and gave a tug at the dress. It didn't help much, except to cause Auntie to look down and back and see what had happened.

At dinner, Aunt Midian caught the girls looking at her. Her face broke into a radiant smile. She clicked her tongue and winked, maybe she was human after all.

FROM CORNCOBS TO CATALOGS

The torture of that icy seat could make a Spartan sob,
For needs must scrape the gooseflesh with a lacerating
cob.

— *J. W. Riley*

James Whitcomb Riley's family privy must have had an outdated supply of corncobs because old timers tell me that fresh ones are not all that uncomfortable. The term: "rough as a cob" could perhaps apply to produce left out in the sun and rain for a year or so, but not to the supply of month-old corncobs used in privy confines. Guests often had a choice of colors, even before the invention of toilet paper. According to privy folklore red cobs outnumbered white ones by a two to one ratio. The modus operandi was to use a red cob first and then follow up with a white one to see if another of the red variety was necessary.

Corncobs have other uses too. They make economical, cool-smoking pipes and excellent quick-start kindling wood for fireplaces and wood stoves. The earliest American farms had extra corn cribs set aside exclusively for shelled corn cobs. These cobs were prized for their oven ash which was used for smoking meat.

From 1888 Hardware Catalog

TOILET FIXTURES.

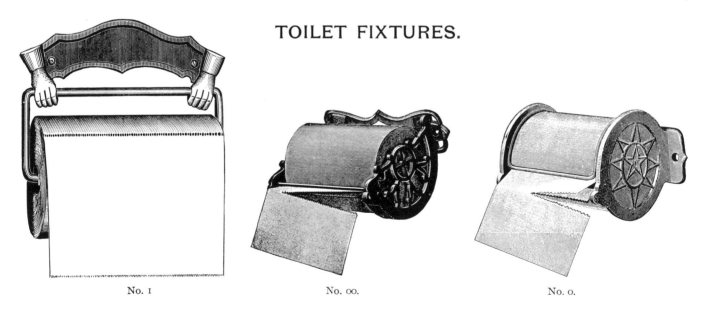

No. 1   No. oo.   No. o.

No. 1, Family, Bronzed, to be Used with Perforated Paper,	per dozen,	$ 2 00
00, Union, " " " Factory or Economy Paper,	"	7 50
0, " " " " " "	"	10 00

TOILET PAPER.

Factory not Perforated,	per dozen rolls,	$2 00
Economy, " "	" "	2 50
Perforated, Best Quality,	" "	3 00

Mail Order Catalogs came into general back house use in the late 1880's. Prior to that time they consisted of less than a dozen pages and could not compete with newspapers, dress patterns and other *uncoated* paper stock. Concerned mothers routinely removed the "female undergarment" and "personal hygiene" sections of these catalogs before consigning them to the outhouse. By the early 1930's most magazines and mail-order catalogs had converted over to slick clay-coated pages and fell into general disuse as a T.P. substitute.

The following letters, reproduced from Bob Sherwood's 1929 book, *Hold Everything,* illustrate how important the thick semi-annual Sears catalogs were to many country households!

Sears, Ward & Co.   Oshkee, Indiana
Chicago, Ill.   July 3, 1928

Gentlemen:
    Please find enclosed money order for one dollar ($1.00) for which please send me ten packages of your Peerless Toilet Paper.
    I'am

                    Yours sincerely,
                    Abner Bewley, Sr.
                    R.F.D.#2 Box 8

(THE REPLY)

Abner Bewley Sr., Esq.   Chicago, Illinois
R.F.D. #2 Box 8   July 6, 1928
Oshkee, Ind.

Dear Sir:
    We acknowledge receipt of your order with enclosure of $1.00 in payment for ten packages of Peerless Toilet paper.
    We assume you have taken this price from one of our old catalogues. On account of the recent increase in the cost of manufacturing this article, the price is now listed at $1.50 for ten packages. On receipt of an additional fifty cents, we will forward at once.

                    Very respectfully,
                    SEARS, WARD & CO.

(THE BACK-FIRE)

Sears, Ward & Co.   Oshkee, Ind.
Chicago, Ill.   July 10th, 1928

Gentlemen:
    I am in receipt of your reply to my letter ordering ten packages of Peerless Toilet Paper.
    If I had had one of your old catalogues, I would not have needed any toilet paper.
    Please send me your latest catalogue, and return my money.
    I am,
                    Yours Sincerely,
                    Abner Bewley, Sr.
                    R.F.D. #2 Box 8

P.S. After thinking the matter over you had better send two catalogues, as we have a very large family of children.

Toilet paper is a fairly modern invention. Today's thoughtfully perforated product was patented by an Englishman named Walter J. Alcock, in the early 1880's. At first there was little if any demand for toilet paper by the roll. British pharmacy owners stocked this item under their counters and out of sight; T.P. was an affront to Victorian sensibilities!

Mr. Alcock was undaunted by the public's reserve and promoted his product religiously. His single-minded missionary zeal eventually paid off. By 1888 toilet paper fixtures (roll-holders) were stocked in most hardware stores, and today Alcock's original factory exports two-ply tissue to a world-wide market.

One can write very easily in pen and ink on modern day Czechoslovakian toilet issue, which is of the consistency of writing paper. My son's college roommate just sent him a long letter on some of this "poor man's stationary". German-made bathroom tissue is light gray in color and rather coarse textured. The brand used on railway trains is imprinted "Deutsche Bundesbahn" on every single sheet. In England, museum-going tourists are quick to notice that each square of paper is plainly marked "Official Government Property".

In some Scandinavian restrooms the extra heavy roll is simply too large to carry away. Mexico solved her paper pilfering problem in airport and bus station "banos" *by not supplying any at all.* Be prepared and bring your own, or you will be forced to borrow from an adjacent stall-holder. Upscale European consumers are switching to a new luxury brand of paper; the pre-moistened, perfumed squares, sealed separately in foil envelopes, are not unlike the "wet wipes" mothers buy in American supermarkets.

Yes, there are enough different varieties of toilet paper to inspire a major collection. (Smithsonian... are you listening?)

What on earth did folks use before printed matter was in wide circulation? Affluent Romans used sponges, wool, and rosewater. The rest of the world grabbed whatever was at hand, including shells, sticks, stones, leaves, hay, or dry bones. Royalty in the Middle Ages was fond of silk or goose feathers (still attached to a pliable neck) for this delicate clean-up task. I'll stop here because this was not intended to be an X-rated book, just use your imagination.

Postscript. According to a 1988 article in *The San Diego Union,* fully 11.5% of all Japanese homes are now equipped with deluxe flush toilets which have built-in hot-water cleansing and hot-air drying mechanisms; these features preclude the need for any sort of tissue at all.

What are the chances of a Japanese-owned electronic-potty-maker locating its manufacturing facilities in San Diego, or some other U.S. city? "Almost nil", stated a spokesman, "Only the quality control standards in Japan are strict enough to produce these devices without fear of public-liability lawsuits."

LOCAL COLOR
by Lewis Cooper

Back in 1955, we bought an 18th-century house on seventy acres in a place named South Galway in Saratoga County, NY. It had five fireplaces but no central heat and very little plumbing or wiring. The first thing I did was to put in an indoor toilet and declare the lovely "three-holer" surplus. We decided that it would help to cement relations with our rural neighbors if we gave away the outhouse.

So we passed the word that it was available. Pretty soon a local citizen, who lived in an establishment we had named "Cockroach Manor", stopped by. He admitted that he lacked such a refinement and indicated interest in taking title to the structure. I suspected that it might be indelicate to inquire into the nature of his present sanitary arrangements; but considering the fact that he had fathered some twelve children by two or three different wives, I figured my offer would solve a rather pressing problem.

On the following Sunday morning, the Squire of Cockroach Manor arrived on the scene with his tractor, hay wagon and three young helpers. They tilted the privy onto the hay wagon but unfortunately neglected to consider the weight of the slate roof, which happened to come down on the side of the wagon which had a slit in one of the tires. The inner tube squeezed out of the tire and exploded, followed by a shrill whistle.

Knowing that the approach of warm weather might make me anxious to cover the hole, they dragged the hay wagon with the outhouse on it about 200 feet out into my orchard, at which point they decided the deflated tire would not survive the trip home. So they tipped the outhouse back upright and departed, saying they would return the following Sunday with another tire.

Bright and early Monday morning the crusty old Vermont farmer, from whom we had originally purchased the place, showed up to claim his hay fork, which he had left hanging in the barn. I was pretty sure that said hay fork had been there when he bought the farm, and that it was legally part of the property (being built into the main roof beam), but I was not going to argue the point.

In our part of the country it is customary to greet neighbors with some trivial comments about the weather, or whatever, before getting down to business — especially when there might be some uncertainty about one's reception. So, when the old boy started out by observing that we had moved the donnicker, I admitted as much, but before I could explain the exact nature of the complex undertaking, which was still in progress, he eyed me sharply and added "Kinda fur to walk, ain't it?" No doubt he had concluded that this fool from the city had moved the privy way out there in order to keep the flies away from his house, and that he would surely regret it come winter.

Although it had nothing to do with the privy project, the old farmer's wife suddenly remarked, with unconscious humor, that they were "getting on" and had health problems, and thought it was time for them to "move back closer to the cemetary". I nodded my head in agreement and went to fetch the hay fork.

Later, the Squire who had purchased our outhouse, confided in me that it was a source of great frustration for him and his present wife to go to the local roadhouse on Saturday night and see his "ex" drinking whiskey with her new boyfriend and paying for it with his alimony check while he had to make do with beer on the more plebian side of the joint. Over the next few years he did various odd jobs for me and was a good worker, but I learned the hard way never to pay him off until a job was finished.

The Squire of Cockroach Manor.

Sturdy Vermont farm stock.

THE ACROBAT

Somehow modesty and pride are intermingled in the human psyche; if you offend one you may also harm the other. Take for example the story of the embarrassed student, adapted from Leon Hale's book *Texas Out Back,* Madrona Press, 1973:

There was a little one room school house at Crabb's Prairie, the teacher there was a man. One day two of the oldest girls in class asked to be excused to go "out back". One of the young ladies was the class show-off and quite a gymnast as well.

In the privacy of the backhouse she demonstrated to her classmate how she could lay on her back and lock her feet behind her head. Well, this time they did not come unlocked, and try as they would, neither student could untangle the rapidly stiffening limbs of the outhouse acrobat.

Finally the distraught witness ran back to the schoolmaster, tearfully imploring his aid lest her friend perish in this undignified position.

After considerable effort the young male teacher freed the red-faced gymnast from her frozen posture on the privy floor and helped her to her feet. Without a word she ran from the outhouse and never returned to school again. Victorian modesty and wounded pride had ended her formal education at the tender age of fifteen.

Modesty is the best policy.

William O. Hickok

William O. Hickok

William O. Hickok

John E. Swartzel

Ulysses S. Grant, 18th president of the United States, used this outhouse at his boyhood home in Georgetown, Ohio from 1823 to 1837.

(near left) Outhouse for Oriental laborers at Meiser's Mill in Snyder County, PA.

(far left) A pristine "house of office" at Old Mill Village in Susquehanna County, Pennsylvania.

(right) Mrs. Bruce Abel had her family's 120 year old outhouse, complete with General Grant cupola, moved to its present backyard location in suburban Cincinnati, Ohio.

(near left) A servant's "necessary house" located at Pennsbury restoration.

(far left) A reconstructed brick "convenience" modeled after the original, owned by William Penn.

Kent & Donna Dannen

(bottom row, both pages) Georgetown, Colorado mining baron, W.A. Hammill, invested part of his fortune in fancy real estate. Hammill House was built in 1879 and featured such luxury appointments as central heating, built-in bathtub, gas lights, glass atrium, and a six-hole outhouse. The cupolaed roof and gingerbread trim harmonize with its lavish interior. The front entrance leads to three solid walnut seats of assorted sizes, which were reserved for family members only. Servants used the back door of the privy which led to a more humble pine plank.

Author's Collection

MOUS PUBLISHING HOUSE FOR POETS IN THE OLD DAYS.

(above) 1940's postcard captured an early California 3-holer.   (below) Four privies share the lens in this 1988 photo of Bodie mining camp.

Frank Pennington

Library of Congress

(above) "Long drop" privy on early farmstead in Lowell, Vermont.   (below) Portable privy for field workers on a Morrisville, Penn. farm.

Department of Agriculture

Vance Packard

Wooden vent identifies this ghost town survivor in Madrid, New Mexico back lot. The town was run by the A.T.S.F. railroad in the 1880's, as a producer of coal. A succession of owners followed until cheap natural gas put them all out of business at the end of WWII.

Most of the company-owned worker's cottages are still standing and lots of relics remain for tourists to browse among. Located a few miles Southwest of Santa Fe on State Highway 10.

C. & M. Shook

Another Bodie, Calif. lean-to. This is the best authentic old gold mining town remaining in the West. In its heyday it supported sixty-five saloons and several bawdy houses. Take old U.S. 395 to Bridgeport, CA and turn east on the dirt road, seven miles south of town. Thirteen very bumpy miles later you will understand why vandals never were much of a problem in the area. Don't attempt the trip in winter or early spring, ten feet of snow covers the 8,000 ft. High Sierra landscape.

George Ford

Cripple Creek, Colorado mining equipment dump features a narrow one-holer fashioned from a used packing crate. Today you can ride the narrow old steam railway on a tour of this former bonanza camp which, between 1890 and 1930, produced a record $400 million in gold.

Cattle pen privy stands a lonely vigil in Leadville, Colorado. First called "Oro City", the played-out camp experienced a name change in 1878. The new boom started when a prospector noticed the dirt he was moving around on his hillside gold claim was twice as heavy as it should be. Analysis proved it out as carbonate of lead, combined with silver. Word of the rich, shallow strike, spread like wildfire and the town grew to 15,000 souls in less than a year! By 1880 Leadville rivaled Denver in population and importance.

Dick Young

Winter snow completely covers these Bodie, Calif. mining camp privies. The buildings here are maintained by the National Park service in a "state of natural decay". There are no conveniences for modern-day tourists, so bring your own lunch and drinking water.

Frank Pennington

Bodie's fire station still stands but this privy's days are numbered. Churches, jails, banks and outhouses are typically the last remaining buildings in a ghost town. The former, because they were well constructed; and the latter, because of their short roof spans.

Paul R. Jones

C. & M. Shook

"Goodbye, God, I'm going to Bodie" was the prayerful lament of a little girl who's family was moving from Colorado to wild and wooly Bodie, California, in the year 1878. With sixty-five saloons and at least one shoot-out a day, the town was known far and wide as the roughest, toughest, mining camp in the West!

The Bodie privy above still retains scraps of its original heavy canvas covering. In the severe High Sierra winters nearly every building in town disappears under a deep blanket of snow. Those few miners, gamblers and ladies of the evening who remianed on the site during the coldest months had to tunnel their way to relief.

Gold miner's privy overhangs Canyon Creek in Burke, Idaho.

Courtesy Norm Weis, author of "The Two-Story Outhouse", Caxton Printers, Ltd. 1988

Creek-drop biffies out numbered regular outhouse installations by ten to one in the once-booming mining district of Burke, Idaho. There are so many privies along the stream above (which runs through the center of town) that locals changed its name from "Canyon Creek" to a four letter designation.

Norm Weis, author of the recently published book "The Two-Story Outhouse", says the creek was an open sewer, outhouses emptied into it from both sides for a mile up stream. There were overhangers, trestle jobs, straddlers, and even a two-holer located smack in the middle of a twenty-foot-wide bridge.

Norman's avid research (which always began at the local watering hole) unearthed many other interesting facts about the area. "Above Burke were the Tiger Poorman Mine, The Hercules, Tamarack, Custer, and Neversweat Mines. Gold had been found in the area about 1860 but large-scale mining did not occur until the 1880's. In 1892, the fights between mine owners and organized mine laborers began.

Dissatisfaction with working conditions continued on for several years, and a serious war broke out in 1899. Over one hundred mine workers commandeered a company train, loaded it with dynamite and forced the engineer to go west to Kellogg. They stopped the train under the bunker Hill Mill and lit the dynamite. Two men were killed in the ensuing battle and the army was called in to restore order. More than twelve hundred miners were arrested!

Later, in my research, I learned that a whorehouse once stood on the creek bank, and a level catwalk extended to an outhouse that stood over Shit Creek on twenty-foot stilts. An older gentleman from Northport told me about it."

"Used to go to work in the morning shift, about four o'clock—still kinda dark. That's when the gals were finished workin' and were doin' their chores out back. We used to applaud each one when they walked across the catwalk back of the cathouse. Sometimes they'd look down and take a bow. We applauded everything."

Ethel Johnson

Back alley convenience is nailed firmly to the adjacent carriage house for security.

Gail Denham

Four old friends in their final resting place at Hunter's Hot Springs Resort in Lakeview, Oregon. Owners, Jim and Vicki Schmit have an extensive outhouse collection and a unique museum.

D. G. Arnold

"Cursum Peregi" (I have finished my course). Abandoned two-holer near Yuma, Colorado.

You work and work, for years and years, you're always on the go. You never take a minute off—too busy making dough. Next year for sure, you'll see the world, you'll really get around. But how far can you travel when you're six feet underground.

An 18th century stucco-covered stone privy at Forge Farm (Mary Ann Iron Works) in Chester County, Pennsylvania. A rare survivor!

William O. Hickok

A fully operable back yard privy in Penn. Dutch country.

These many folk structures are of the soil. They are natural, though often slight; their virtue is intimately related to the environment and to the heart-life of the people.

—Frank Lloyd Wright

William O. Hickok

A well ventilated colonial period "necessary house" located at Williamsburg Virginia. A full time museum staff of 2,886 personel provide major public education programs and costumed workers and guides at the famous historic restoration. Collections represented date from 1492 to 1865.

William O. Hickok

William O. Hickok

Michael Arford

Bucks County, Penn. backhouse. Note hex sign on ventilator.

Birdhouses were privy status symbols often set on rooftop poles.

Typical National Park "comfort station" design of 1920-1950.

This Tortilla Flat, Ariz. landmark appears on many postcards.

Mary Okey

Susan I. Davison

Berman E. Ross

Victorian public park privy, (panelled door has been covered).

Richard Longseth

Alaska has more outhouses than any other state in the Union.

Stately "house of office" stands behind 1870's brick mansion.

Hats on posts signal that this New Mexico privy is occupied.

Lynn R. Fox

Gail Denham

Drunk or sober, patrons had to "walk the plank" to reach this outhouse located behind a barroom in Pilottown, LA.

Library of Congress

"Long Drop" used by poor folks living in shacks along the highway between Charleston and Gauley bridge. Sept. 1938
— *Marion Post Wolcott*

Pemaquid, Maine. Shingle covered privy still stands behind a 1792 white clapboard home on stone foundation.

Half a boat is better than none. Outhouse conversion near North Rustico Beach.

Victorian two-holer sits on edge of freshly plowed field. Roof brackets match those of original farm house.

*"All the places that the eye of heaven visits,
Are to a wise man ports and happy havens."*

—Shakespeare

*"But when the crust was on the snow
and sullen skies were gray,
In sooth the building was no place
where one could wish to stay."*
— J. Whitcomb Riley

*"Remove not the ancient landmark which
thy father hath set up."* Abandoned farmstead, Southeastern Ohio, 1959.

*Coast Guard "convenience" located on
Michipicoten Island in Lake Superior.
Buttercups and Hawkweed flourish in fore-
ground.*

*When a man reproached him for going into
unclean places he said, "The sun too,
penetrates into privies, but is not polluted
by them."*
—Diogenes

John E. Swartzel

Plywood door and pre-grooved panelling probably date this backyard biffy from the 1950's. Once a busy three-holer, it now serves as a woodshed.

John E. Swartzel

No, this is not a moonshiner's hideout; it's a public accommodation located along side of the Little Miami River near Waynesville, Ohio.

Ethel Johnson

Ulrich K. Tutch

"Bump-out" style privies were conveniently attached to a barn.

The last double-decker in Cedar Lake, Michigan is 100 yrs. old.

Transplanted Colorado privy now serves as an attractive tool shed.

Part of the outhouse collection at Hunter's Resort, Lakeview, Ore.

D. G. Arnold

Gail Denham

This outhouse was the recent victim of an aluminum siding salesman.

Star cut-out indicates "For Men Only". Privy has plaster interior.

Gunston Hall, Virginia. An accurate re-creation of a pilgrim privy.

New Orleans two-door is exact reproduction of 1850 backhouse.

John E. Swartzel

Doctor's mansion and surgical area, circa 1820. The good doctor was apparently oblivious to any danger posed by the attached privy.

Two of the most dreaded diseases in the past were typhoid fever and cholera. Only the Bubonic Plague (Black Death), which destroyed upwards of fifty percent of the population of England in the fourteeneth century, was feared to the same degree. Unlike the plague, which was carried by rats and communicated to man by fleas, typhoid and cholera were spread by milk and drinking water contaminated by fecal bacteria. Food handled by human carriers, or fouled by flies, was also a culprit. Cholera could claim a victim by fever or dehydration within a couple of hours, but the normal period between contact and severe symptoms was three to five days. Typhoid could take up to three weeks to manifest its presence.

Probably no epidemic in American history better illustrates the awful results of a single thoughtless act than the outbreak of typhoid fever at Plymouth, Pennsylvania, in 1885. In January and February of that year the night discharges of a typhoid patient were thrown out of a window on to the snow below. The germs from this waste matter were carried by the spring thaw into a nearby stream and on into the public water supply. The ensuing epidemic lasted from April to September and a total of 1,104 persons came down with typhoid fever; of these, one hundred and four were fatal cases.

Asiatic cholera was a "foreign" disease, imported from India to England and on to the United States. It usually ran in epidemic proportions and in highly populated areas. From 1832 to 1859 several outbreaks occurred in our midwestern states. The press reported that "People are dying like flies", and St. Louis health authorities were overwhelmed with the six hundred cholera deaths which occured in a single week during 1833. It is recorded that steamboats making their way down the Mississippi river stopped daily to put their dead ashore for burial.

In 1853 an outbreak of cholera in Westminster, England claimed over two thousand lives. The famous Dr. Snow was called in to try to find the cause of the epidemic which had been attribued to "foul smelling air" from decaying vegetation or sewer gas. The prevalent preventative measures ran from burning bonfires on street corners, to individual dosages of calomel, opium, castor oil, powdered mustard, camphor and red pepper. None of these "cures" were effective. Many probably hastened the victim's demise through additional dehydration. (Actually, a high liquid intake would have been the wisest choice to follow.)

In any case, the good doctor set about surveying the parish neighborhood to try and isolate a common culprit. He quickly determined that most all the sufferers were obtaining their drinking water from a public pump located on Broad street. The well there had a good reputation for water of uncommon sweetness and freshness.

A test of the Broad street water supply showed it to be contaminated with organic impurities which had gained access through a neighboring house sewer into which the discharges of a patient suffering from severe diarrhea had been dumped. The outbreak was squelched within a week simply by removing the pump's handle. Those who had routinely purchased bottled water from mountain sources had not been affected by the disease even though they lived side-by-side with the victims. Others were not so fortunate. One mother, who had come to the aid of her ailing married daughter, had inadvertantly touched some soiled bedsheets during the wash day. As she walked home that night she suddenly dropped dead in the middle of the roadway.

John E. Swartzel

This Waynesville, Ohio landmark was a combination dwelling and commercial creamery at the turn-of-the-century. Farmers drove their wagons through the overhanging arbor and unloaded cream cans directly into a basement butter-processing and storage area.

Eventually vaccines were developed and the public partially educated; however, even as late as 1920 it was estimated that twenty-five percent of all U.S. farms had contaminatd water supplies. Drainage from surface slops, animal pens, and poorly situated privies reached many wells through underground seepage. (In some parts of the South, flies simply carried the comma-shaped bacteria from flimsy, open-backed, servant's privies directly to the master's formal dinnertable.)

The French chemist, Louis Pasteur (1822-95), discovered that *germs,* not gas, caused infections; and that they could be killed by heating liquid to a temperature of 145° F. and sustaining it at that point for thirty minutes. The process, followed by quick chilling, was called pasteurization. Mr. Pasteur and modern plumbing methods have ridded western society of cholera and typhoid fever but undeveloped nations still have a high death rate from these preventable diseases.

THE GENTEEL ART OF PRIVY DIGGING

O ne of recent history's best kept treasure hunting secrets is privy hole digging. Who could have imagined what priceless finds lay at the bottoms of these humble refuse pits? To put it in a more practical perspective, think of back house vaults as time capsules of material culture. The difference between privy-encased records and those found in office building cornerstones is that in the latter objects are chosen by committee. Contributions to privy pits were much more spontaneous.

Outhouse excavators often share their "finds" in regional newsletters. This excerpt by Tim Wolter is from the North Star Bottle Club's bulletin.

Since the privy was also a convenient place to dump any unwanted objects some surprisingly unusual items have been recovered. Bicycles, ice skates, a gramaphone, Model T parts, animal skeletons, shotgun shells, blackboards, piggy banks, paint brushes, shovels, garden tools, statuary, umbrellas, sets of false teeth, and even a porcelain toilet from the 1890's are among the items I have dug. Once upon a time we dug one filled with billiard balls. Yet another was full of carpet salesman's sample books. We have also encountered sawdust, tin cans, birch bark and sacks of plaster.

Another dedicated group of earth-movers call themselves the National Privy Diggers Association. Their newsletter, *PRIVY*, is available for ten dollars a year from the editor: Don Dzuro, at 3532 Copley Road, Akron, Ohio 44321. Don recently sent me a large packet of photos and back issues. Among them was this account of a 1986 outing by Association president, Richard L. Wilcox of Mechanicsville, Virginia.

BOTTLE DIGGING IN OLD NEW HAVEN

On July 4th, to celebrate the 210th anniversary of our country, Tim Dolmont and John Grant went out looking for a privy that would have been active before the centennial celebration. Unfortunately they didn't find one quite that old, but they did dig a woodliner, 5' x 4' x 6' deep. The house dated from about 1860 but the bottles were mostly 1880 vintage. Forty whole bottles were recovered.

The "keepers" included an olive Allenville Glassworks whiskey, two aqua cone inks, a paneled ink, an aqua Kickapoo Indian Oil, a square cobalt medicine, a very small clock-shaped bottle embossed Gerstendorfer Bros. and various druggist and perfume bottles. The broken pieces included a honey-amber whiskey, a blue milk glass vase and a china soap dish with a very ornate top consisting of a reclining lady, minus her head and right arm.

On July 12th, the gang went out to dig during a light rain shower. The area they probed was fifteen by twenty feet. The most conspicuous feature of this yard was a nauseating stench that filled the air. Theories about fresh privies and dead bodies in the bushes were finally dispelled by the discovery of an old refrigerator on the porch. This abandoned cooler had an undefined substance dripping from it. At any rate, we found the privy vault which measured 4' x 4' x 4'. Whole bottles included two cone-shaped inks, a salt and pepper shaker ornately embossed with human faces and a black glass ale with a crown on the shoulder.

This dig grew soggy and stayed that way when the shower turned into a soaking rain. (Note that most privy digs are dry and relatively odorless).

Courtesy National Privy Diggers Association

Careless children were frequent contributors to these historical repositories. Everything from marbles to china-head dolls and cast iron toys has been unearthed by diligent diggers. Many items found their way into back house vaults in the form of fill material. Since all but the deepest privies were moved every five or ten years it was necessary to top-off uncovered holes with a solid substance to prevent accidents. Fill matter was often a mixture of ashes, household trash, sand, stones and dirt.

The best method of approaching property owners for permission to dig is to guarantee not to disturb any plants or shrubs; offer them first pick of the booty, and tell them you will be sure to fill in any holes before leaving. Old maps, foundations, fences and surveyors stakes will help you to decide where to begin searching for the telltale clink of glass against your steel probing rod. Remember that back houses were just that — way out back near the alley, barn, cornfield or property line. Happy hunting!

Antique bottle collecting is said to have begun in earnest in about 1959 when a group of California enthusiasts met to form the Antique Bottle Collector's Association and began to circulate their newsletter and recruit members on a national scale. By 1970 there were over one-hundred clubs with thousands of avid members. In the same year Cecil Munsey, a San Diego resident, wrote the first definitive book on the hobby, *The Illustrated Guide to Collecting Bottles* which was published by Hawthorn Books and sold over 250,000 copies.

Obviously bottle collecting had become one of the hottest hobbies in the country, something akin to the baseball card fad of today. Interest is still very strong among many devotees; witness the single bright green "Dr. Wynkoop's Katharismic Honduras Sarasparilla" which brought $7,250 at a recent Bolton, Massachusetts antique bottle auction.

Jim Simmons

(opposite page) Three privy-hole-diggers communicate with a cohort seven feet below the surface.

(at right) More exciting than King Tut's Tomb. After breaking through a concrete slab floor, the partners cleaned out a 5' x 9' x 8' deep brick-lined vault. Many collectible bottles were recovered.

Portrait of a happy PhD. Scott Garrow, of Lombard, Illinois, earned his doctorate by submitting this photograhpic proof of advanced study to the National Privy Diggers Association.

(below, right) Elton Tanner displays part of a treasure trove discovered at the bottom of a seven foot deep excavation.

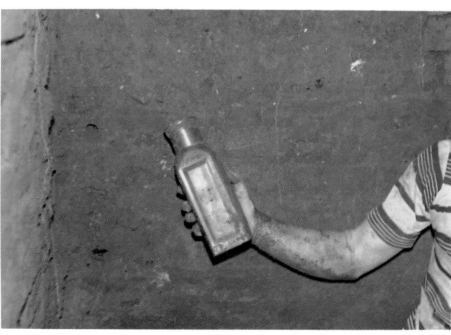

The Three Bears

Ma tried to wash her garden slacks
  but couldn't get 'em clean
And so she thought she'd soak 'em
  in a bucket o' benzine.
It worked all right. She wrung 'em out
  then wondered what she'd do
With that bucket-load
  of high explosive residue.

She knew that it was dangerous
  to scatter it around,
For grandpa liked to throw his
  lighted matches on the ground.
Somehow she didn't dare to pour
  it down the kitchen sink,
And what the heck to do with it,
  poor Ma just couldn't think.

Then Nature seemed to give a clue,
  as down the garden lot
She spied the edifice that graced
  a solitary spot,
Their "Palace of Necessity",
  the family joy and pride,
Enshrined in morning-glory vine,
  with graded seats inside;

Jest like that cabin Goldylocks
  found occupied by three,
But in this case B-E-A-R
  was spelled B-A-R-E.
A tiny seat for baby BARE
  a medium one for Ma,
A full sized section sacred to
  the BARE of old Grandpa.

Well Ma was mighty glad to get
  that worry off her mind,
And hefting up that bucket
  so combustibly inclined,
She hurried down the garden
  to that refuge so discreet,
And dumped the liquid menace
  safely through the center seat.

Next morning old grandpa arose;
  he made a hearty meal,
And sniffed the air and said "By Gosh,
  How full of beans I feel.
Darned if I ain't as fresh as paint;
  my joy will be complete
With jest a quiet session
  on my usual morning seat;
To smoke me pipe an' meditate,
  and maybe write a pome,

For that's the time when bits o' rhyme
  gits jiggin in me dome."

He sat down on that special seat
  slicked shiny by his age,
And looking like Walt Whitman,
  jest a silver whiskered sage,
He filled his corncob to the brim
  and tapped it snuggly down
And chuckled: "Of a perfect day
  I reckon this the crown."
He lit the weed, it soothed his need,
  it was so soft and sweet:
And then he dropped the lighted match
  clean through the middle seat.

His little grandchild, Rosyleen,
  cried from the kitchen door:
"Oh Ma, come quick; there's sompin wrong;
  I heard a dreffel roar;
Oh Ma, I see a sheet of flame;
  it's rising high and higher . . .
Oh Mummy dear, I sadly fear
  our comfort-cot's caught fire."

Poor Ma was thrilled with horror
  at them words o' Rosyleen.
She thought of Grandpa's matches
  and that bucket of benzine;
So down the garden geared on high,
  she ran with all her power,
For regolar was Grandpa,
  and she knew it was his hour.
Then graspin' gaspin' Rosyleen
  she peered into the fire.
A roarin' soarin' furnace now,
  per chance old Grandpa's funeral pyre . . .

But as them twain expressed their pain
  they heard a hearty cheer,
Behold the old raspcallion squattin'
  in the duck-pond near.
His silver whiskers singed away,
  a gosh-almighty wreck,
With half a yard o' toilet seat
  entwined about his neck . . .

He cried: "Say folks, oh did ye hear
  the big blowout I made?
It scared me stiff—I hope you-uns
  was not too much afraid?
But now I best be crawlin' out
  o' this dog-gasted wet . . .
For what I hope to figger out is
  —what the heck I et."

—Robert W. Service

Reprinted by permission of the Putman Publishing Group,

from "Songs of a Sun Lover", © 1949 by Dodd, Mead & Co.

Author's Collection

(above) New Mexico guest ranch privy, circa 1930.   *(below) 1890's fairgrounds facility for females only.*

Bettman Archive

Magazine and postcard artists enjoyed a fifty year love affair with the outhouse begining in 1901 when tiny cardboard privies were given away as souvenirs at the Pan American Exposition. From then until the early 1950's commercial illustrators recorded every conceivable comic situation an outhouse occupant might endure. Today's tourists are still sending home privy art in the form of postcards, bookends, letter boxes, chess sets, salt and pepper shakers, collector plates, and toilet paper holders.

"If you would read the labored scribblings of a black adulterer, seek out the poet with a charcoal piece who writes verses to be read by those seated on a privy." — Vesuvius, privy chamber inscription (79 A.D.).

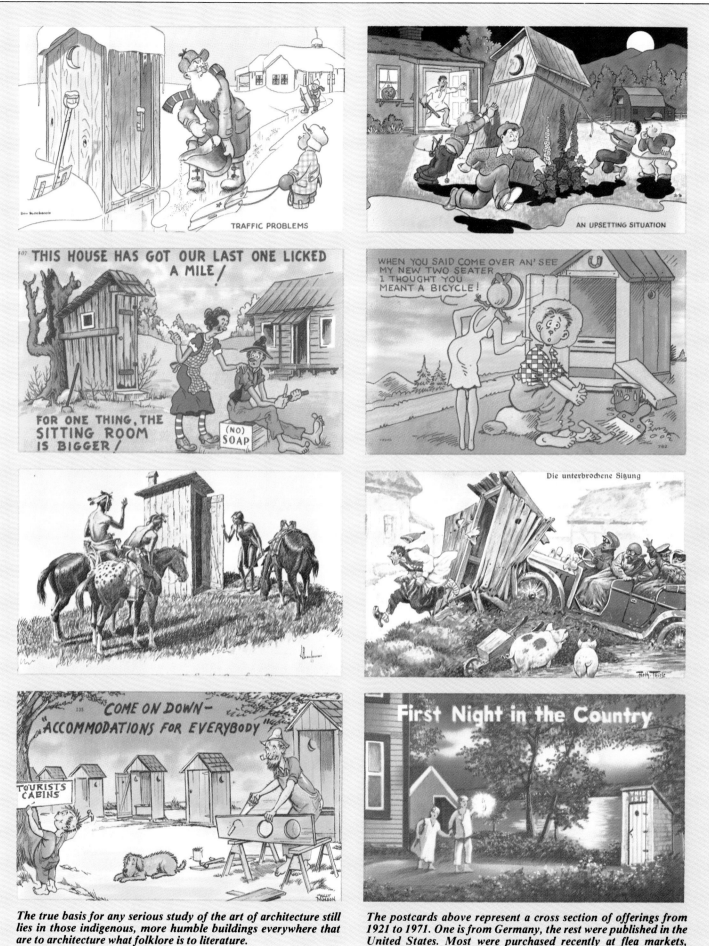

The true basis for any serious study of the art of architecture still lies in those indigenous, more humble buildings everywhere that are to architecture what folklore is to literature.
— *Frank Lloyd Wright*

The postcards above represent a cross section of offerings from 1921 to 1971. One is from Germany, the rest were published in the United States. Most were purchased recently at flea markets, antique shops, and postcard shows for less than two dollars. The oldest, Victorian era cards, are very scarce and command from ten to twenty dollars apiece. An advanced collector, specializing in "privies only" wrote to inform us that he has accumulated over four hundred and twenty-five distinctly different designs.

THE SATURDAY NIGHT BATH

To the Norsemen of old we owe more than just the names of weekdays — Wednesday, *Woden's Day,* Thursday, *Thor's Day* and so on. The Saturday night bath ritual is said to have originated among these robust seafaring warriors.

Their bathroom (wash hut) was built of logs and made airtight with clay except for a smoke hole in the roof. After piles of stones were heated by a fire the whole family assembled inside. The roof opening was then sealed, cold water was thrown upon the hot rocks, and behold... a steam bath. After switching each other with wet twigs to increase skin circulation these crazies ran amuk in the freezing snow. No wonder they were highly feared by much of the unwashed world!

From ancient Greece to the Roman Empire and on through the eighteenth century, bathing was a common public activity — sometimes segregated, sometimes mixed — often a family event.

"A bath in every bedroom" is not entirely a modern concept. Archeologists who dug up Pharoah Ramses' 1500 B.C. palace discovered custom built one-bedroom, one-bath, apartments for each of the concubines in his harem.

This was peanuts compared to the five-acre palace of King Minos who ruled the city state of Knossis on the island of Crete, two-thousand years before the birth of Christ. His indoor tub was covered with figures of flagrant nymphs and the nearby flush toilets were vented to exhaust sewer gases.

Agrippa introduced public baths to Rome in 23 B.C. The baths of Diocletian were the largest and could accomodate over three thousand frolicking citizens at one time. Also provided were two hundred and sixty marble-seated commodes – probably in an effort to keep those acres of bath water crystal clear.

Upper class Romans had private bathing facilities in their homes. These baths often occupied an entire marble-lined room and contained piped-in hot water and heated air.

After the decline and fall of the Roman Empire, sanitation took a back seat to survival for most folks. Early Christians strongly associated the Roman penchant for twice-daily bathing with similar passions of unseemly nature, and it was a very long time before cleanliness ever became associated with godliness.

Several centuries elapsed before royalty began to rediscover "total" bathing. Catherine the Great of Russia had a tin tub installed in a tiny corner of her one-thousand-room palace, and Queen Isabella of Spain even took a couple of baths before she died.

In 1399, Henry the Fourth of England founded a military order called the Knights of the Bath. Part of the initiation ceremony included a shave, a haircut, instruction in the laws of chivalry, and a ritual bath.

Crusaders, returning home from Palestine, brought with them new ideas of personal hygiene. Their enemies took frequent baths and were not subject to the recurring plagues of Europe where the Black Death had destroyed one-fourth of the population.

Queen Elizabeth I ordered a bath built inside her royal chambers and the court scribe duly recorded "She doth bathe herself once a month whether she require it or no."

Later, the great unwashed gentry of Britain actually learned to bathe from their Hindu subjects; although it did take several generations for the daily bath to become an established institution in England.

In early America the bathroom was of course nonexistent. Bathless homes were a normal state-of-being. Washing your face, neck, arms and feet once a week was considered adequate for general social acceptance. Special containers were kept handy for this chore in upper class bedrooms, where bathing customarily took place. These furnishings consisted of a washstand, a bowl, a pitcher, a slop jar (for used water), a foot bath, and a towel horse. Warm water was carried upstairs by servants; plumbing in bedrooms was a late nineteenth century invention.

TOILET SETS.

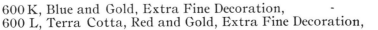

No. 600 K.

600 K, Blue and Gold, Extra Fine Decoration,   - - - - - per set, $5 00
600 L, Terra Cotta, Red and Gold, Extra Fine Decoration,   - - - - " 5 00

MOTT'S PATENT BATH AND SHOWER COMBINATION.

BATH AND SHOWER COMBINATION.

COMPRISING

The "Imperial" Porcelain, Porcelain-lined Iron or Planished Copper Bath, with Supply and Waste Fittings and Patent Shower and Curtain.

THE Patent Shower and Curtain may be used in conjunction with our "Imperial" Porcelain, Porcelain-lined Iron or Planished Copper Baths when they are encased as illustrated above or when they are set up open; in the latter case the Polished or Nickel-plated Brass Pipes run up and are attached to the tile or wood wainscoted wall. When the Shower is to be used the Curtain is removed from the hook and envelops the bather, who can have a perfect shower bath without splashing the cabinet work, wall or floor.

When the two lower valves, lettered "Hot" and "Cold," are opened, the water passes into the mixing column, attached to which is a thermometer; the latter registers the degree of temperature of the water. From the mixing column the water passes to the shower by opening the upper valve.

J. L. Mott Catalog of 1888

Most folks cooked up their own lye soap in a kettle on the back porch. But for those who could afford store-bought goods, Pears' soap was available in every city in the world.

The bath water recipes are from a popular book of household hints published in 1908.

People still believed a good hot bath could cure anything.

PEARS'
SOAP
a·Specialty for·Children

Salt-water Bath. — Add 4 or 5 pounds of sea salt, which can be purchased of any druggist, to a full bath at the temperature of 65° F. The patient should remain in this bath from 10 to 20 minutes, and afterwards should rest for half an hour in a recumbent position.

Mustard Bath.—The addition of 3 or 4 tablespoonfuls of powdered mustard to a hot footbath in cases of chill is a preventive against taking cold, and is also useful in the early stages of colds to induce perspiration. The feet should be taken out of this bath as soon as the skin reddens and begins to smart. The parts bathed should be carefully cleansed, rinsed, and wiped dry. Great care should be exercised in giving mustard baths to children, else the skin may become badly blistered.

The Bran Bath.—Make a decoction of wheat bran by boiling 4 or 5 pounds of wheat bran in a linen bag. The juice extracted, and also the bran itself, should be put into the water. This is for a full bath at a temperature of about 90° F. This bath is of service in all skin affections accompanied by itching.

Benjamin Franklin brought home his own shoe-shaped bathtub from Paris in 1778. He was suffering from a skin ailment at the time and took hot baths often. He sat erect in the heel with his legs extending into the vamp of this giant shoe. There was a device in the heel to keep the water warm and the toe contained a spigot drain for emptying. A book rack was built into the tub at eye level for leisurely reading. The entire contraption was made of copper for easy maintenance.

Seventy-two years elapsed from the time of Ben Franklin's pioneering efforts until a bathtub was actually installed in the White House, by President Millard Fillmore, in 1850. Sometime earlier the state of Virginia had placed an annual tax of thirty dollars on every bathtub brought into the state, and Boston had made bathing unlawful except on medical advice. President Buchanan even lacked a tub at his private residence in the late 1860's.

Why all of this commotion over bathtubs? Well modesty we suppose — probably a puritanical heritage of sorts. It is interesting to note that in early America, bathtubs were considered highly objectionable to look upon — much like a urinal would be to some folks today. Slipcovers and wooden cabinets were made to hide them from public view when not in use.

Gradually, in the 19th-century, all this changed — "Cleanliness" had become "Godliness" — dirt and bodily filth were condemned by all right minded persons. The Saturday night bath became almost a religious ritual.., you had to clean up your act before Sunday morning. Church services never smelled better!

During the 1870's, when all plumbing began to move indoors, cast iron bathtubs gradually replaced the earlier portable tin variety which were dragged out for use in front of the fireplace or woodstove on Saturday night. It had taken four thousand years to come full circle. Bathers were back in the tub where they belonged!

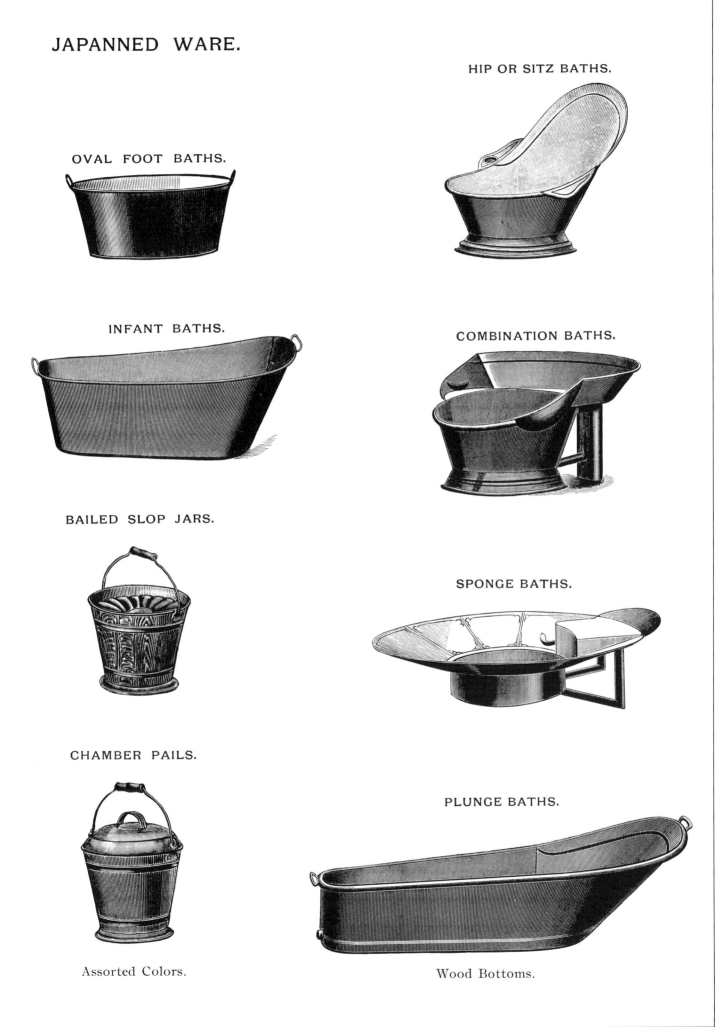

JAPANNED WARE.

HIP OR SITZ BATHS.

OVAL FOOT BATHS.

INFANT BATHS.

COMBINATION BATHS.

BAILED SLOP JARS.

SPONGE BATHS.

CHAMBER PAILS.

PLUNGE BATHS.

Assorted Colors.

Wood Bottoms.

1877 Advertisement

From 1888 J. L. Mott Iron Works Catalog

VICTORIAN BATHROOMS

Americans were late comers to toilet training. Although basic indoor fixtures had been available since the end of The Civil War we must remember that five million folks, about 64 percent of the work force, still lived down on the farm. Most of them had never seen a real flush-toilet.

Many city dwellers were not much more refined than their country cousins. Since the early 1800's it had been the custom in some towns to dump household slops and chamber pot residue directly into open street sewers where grazing porkers converted it into breakfast table fare for less affluent residents.

As the unwashed masses gradually became more aware of the benefits of daily bathing and indoor waste disposal, manufacturers rushed to fill an increasing demand for fancy plumbing fixtures.

The 1888 catalog of the J. L. Mott Iron Works, of New York, featured over 275 lavishly illustrated pages replete with everything from dolphin shaped commodes to clawfoot bathtubs and imported French bidets.

The enameled china interior and exterior surfaces of toilet bowls provided designers with a field day. Their gaudiness knew no bounds. A popular magazine writer of the day declared of the vogue for garish fixtures,

"The stars and sky, and all the zodiac of heaven shone up at me from that bowl upon which I sat. And when I pulled the handle I watched with awe as the whole surged to the brink of the milky way and was gone.

After World War I there was a huge building boom in the United States. Manufacturers were riding the crest of an economic wave brought about by exporters flooding war-ravaged Europe with American-made goods. The standard five by seven foot bathroom found in most middle class homes of today, dates its inception from this period. In the 1920's outhouses were banned in all urban areas and laws were passed requiring the installation of a bathroom *complete with toilet* in every new dwelling constructed. Supply was no problem. Sears & Roebuck stood ready to ship a modern water closet including an instruction booklet, a copper-lined golden-oak tank, and a contoured hardwood seat, for $11.95.

Pull-chain models were less expensive but were loud enough to wake the whole neighborhood. Water came rushing down a shiny brass pipe and entered the bowl with a gratifying roar! Kids never tired of these indoor waterfalls.

MOULE'S PATENT EARTH-CLOSET

The simplest form in which the earth system can be applied is to have at hand a box of sifted dry earth, with a small scoop or tin cup with which to distribute it over the feces, or to keep the supply in small paper bags, each holding enough for a single operation, say a little more than a pint. This will afford a satisfactory means for testing without cost the efficiency of the application; but it will lead all who care for convenience and nicety to the subsequent adoption of the Rev. Mr. Moule's mechanical system, or its equivalent.

Price $20

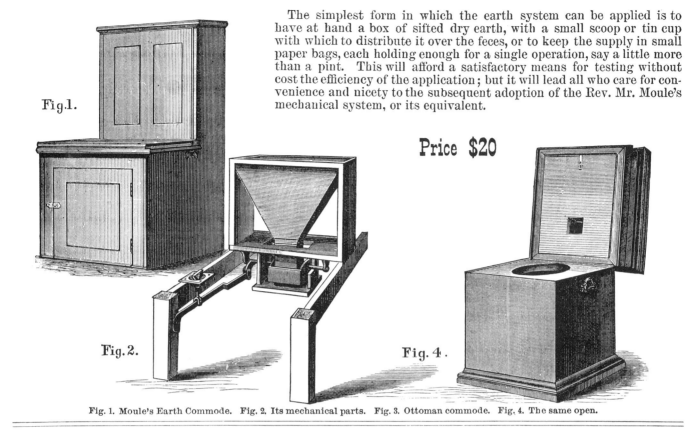

Fig. 1. Moule's Earth Commode. Fig. 2. Its mechanical parts. Fig. 3. Ottoman commode. Fig. 4. The same open.

In the year 1860 Rev. Henry Moule rediscovered *and patented* a principle which house cat owners have known for thousands of years. His marvelous "discovery" was the fact that fine dry earth, or sifted stove ashes, can render the products of human elimination odorless, inoffensive, and almost invisible. You see, normal night soil is over seventy-seven percent water, and moisture is very quickly absorbed and dispersed by dry sand. If you use a bucket full of dry earth, or ashes, per movement you can render the substance almost harmless in a very short time. The advertised payoff was instant garden-ready fertilizer, no plumbing pipes, no wasted water, and an odor-free parlor when company came (and went).

Reverend Moule, Vicar of a Dorsetshire, England parish, made waves on both sides of the Atlantic with his new portable potty, and imitators sprang up like weeds! The U.S. Department of Agriculture wrote up his invention in its annual report of 1871 and an additional two-hundred-fifty-thousand extra copies were circulated privately. According to this thirty page dissertation, an average human adult produces two pounds of feces daily. Although mostly water by volume, it also contains urea, slime, sulphate of potash, sulphate of soda, uric acid, phosphate of ammonia, muriate of ammonia, phosphate of talc and lime, silic acid, carbon, hydrogen, nitrogen, and oxygen.

When scientists compared it to the best brand of commercial bat guano they discovered that an average American's waste matter was about one-third as potent. Seven hundred thirty pounds, a year's output, was worth ten dollars in gold in 1871.

Henry Moule's "Earth Closet" did not make a significant dent in the domestic dooley market. The American version was priced at twenty dollars plus freight, more than a month's wages for most folks. Household help was already spread too thin to take on the added burden of carting twenty pounds of sanitized sand up and down stairs twice daily. The cost of the system simply outweighed its benefits.

Digging a deep hole in the ground and moving the family privy a few yards south every five or ten years, remained the state-of-the-art in rural plumbing for another half century.

Just think, if Rev. Moule had been as successful in the waste disposal business as his fellow countryman, Sir Thomas Crapper, we might be using the word "moule" in place of another common synonym.

Sir Thomas Crapper was a pioneer in the refinement of all sorts of porcelain conveniences. In the 1880's Sir Thomas (he was knighted for his efforts) was commissioned to install modern biffies, bidets and urinals in several dwellings of the royal family. A decade later he was still coming up with new ideas, and in 1891 was granted a patent for a revolutionary seat-activated flushing device.

Mr. Crapper's main business however, was the manufacture of water closets. T. CRAPPER & CO. was conspicuously emblazoned upon every one of the thousands of potties that left his warehouses. The name gradually became a generic term for toilets, and our impressionable young troops brought it home from Europe at the end of World War I.

BIBLIOGRAPHY

RURAL LANDMARKS
by William Colson
If you ever advertise as a cash buyer for books about outhouses, this is the title you will be offered most often. I can't imagine who thought this sixty page volume very amusing, even in 1945, but apparently several thousand copies were sold.

A sophomoric wordplay on outhouse and bathroom synonyms replete with old laxative ads, silly cartoons and corny newspaper clippings. Whatever happened to the Highlights Publishing Company, of West 99th Street, Chicago, 43, Illinois?

PRIVY, OUTHOUSE, BACKHOUSE, JOHN
by Wellington Durst
Fifty-two pages of interesting insight by someone who obviously lived much of his life in old back house days. Here is a sample:

Question: We know that early jail houses had privies out back like other buildings of the time. Was there a guard on duty twenty-four hours a day to escort prisoners to the John?

Answer: Yes, they had what they called sheriff's assistants. When a prisoner had to "go" he would call out, "Take me to de potty." These special assistants would escort them out and bring them back. Eventually they became known as "De Potty Sheriffs" which evolved into the 'Deputy Sheriffs' we have today. Copyrighted in 1980 by the Watne Fogle Company. Grove City, Pennsylvania 16127.

THE DONNICKER BUILDING BOOM
by Newton Easling
The author chose to close his business doors rather knuckle under to New Deal politicians who were squandering his tax dollars on public welfare projects. This sixty-two page satire of the Works Progress Administration was Mr. Easling's way of getting back at the incumbent Democratic administration and hopefully swaying the outcome of the upcoming presidential election. Full of Donnicker photos and cartoons. Well worth the fifty cents a copy charged in 1938. Published in Pekin, Illinois, by the author.

EN BOK OM AUTRADEN
(A Book About Outhouses)
is the title of this Swedish volume which is filled to the brim with wonderful privy facts and photographs. We don't own a copy but were told that it is still in print.
Published by Ferrosan, Box 839, 20180, Malmo, Sweden.

BACKYARD CLASSIC
by Lambert Florin
Ghost town depositories are the overiding theme of this classic contribution to outhouse and mining town lore. The author and his buddies scoured desert floors and mountain peaks to round up enough delapidated cabins and comfort stations to fill this large format, 160-page, photo collection. First copyrighted in 1975 by Superior Publishing Company of Seattle, Washington.

TROPHIES OF AMERICAN INGENUITY
by Lynn Fox
Lynn Fox has been sketching, shooting and painting outhouses for more than twenty-seven years. This little fourteen page booklet, printed by an advertising specialty firm in 1985, is a pretty good sampler of his work. Mr. Fox draws a Sunday comic strip entitled "Johnny" for the Dover New Philadelphia Reporter and has also issued a privy collector plate. The booklet is available for two dollars postpaid from the author at 240 W. Main, Carrollton, Ohio 44615.

OLD GEORGIA PRIVIES
by Mary Frazier and Dean Long
A thirty-nine page selection of sepia-toned Georgia privy photos taken on weekend outings over a six year period. Mrs. Long, a fifth grade teacher, spent eight months choosing appropriate poems and quotations for the captions. You can still purchase a 1984 edition at the local drug store in Gwinnett or send six dollars to Mary and Dean Long at 288 Craig Dr., Lawrenceville, GA 30245.

GEMS OF AMERICAN ARCHITECTURE
by William R. Greer
Even back in 1939 the calendar people, Brown & Bigelow, realized what a great advertising attraction candid photos of privies might be. So they put together a pocket-sized fifty-two page collection of outdoor conveniences with clever captions and space for a client's imprint on the back cover. Evidently millions of these little pamphlets were given away by insurance agents, automobile dealerships, hardware stores, etc. You can pick up a copy from an antiquarian book dealer for about ten dollars.

TEXAS OUT BACK
by Leon Hale & Harvey De Young
Author Leon Hale's hobby is visiting all the roadside cafes and honky-tonks he can work into his busy schedule — and you can bet that he has picked up some very interesting anecdotes along the way. In 1973, Madrona Press of Austin, Texas asked Leon to write an introduction to a portfolio of original privy pencil sketches by San Antonio landscape artist Harvey De Young, who died in 1956. These drawings are the most sensitive treatment of the subject I have ever seen and Mr. Hale's narrative is as smooth as roadhouse gin. Try contacting used book stores in the San Antonio area.

METAMORPHOSIS OF AJAX
by Sir John Harington
The first book ever published on the subject of water closets was this tongue in cheek instruction manual written in 1596 by John Harington, godson of Queen Elizabeth I. Sir John was the inventor of England's first flush-toilet, and the good queen, being a practical woman, had the book bound in wooden boards and chained securely to the wall of her closet chamber. There is a distinct possibility that the polite euphemism "John", evolved from Sir John's widespread notoriety and the poetry he wrote about water closets.

OUTHOUSES OF THE ALLEGHENY HILLS
by Dwight Hayes
The librarian at Hinkle Memorial Library in Alfred, New York, told us that this forty-five page book is a very good photographic record of early Allegheny Mountain privies. It was published by the Misty Lantern Studio, Arcade, New York in 1971.

THE PASSING OF BACKHOUSE BILL
by E. Herron and L. Ordway
Dedicated to Fred K. Ordway, "Alaska's flying photographer, whose camera skill uncovered the humor and beauty of the great Northland". A home-grown thirty-two page booklet of privy photos and sourdough prose, published and sold for sixty cents in 1939 from Box 2511, Juneau, Alaska.

OUTHOUSES OF THE EAST
by Sherman Hines and Ray Guy
Possibly the most beautiful collection of Canadian privy photos ever assembled, and certainly part of our inspiration for publishing *The Vanishing American Outhouse*. This handsome seventy-one page hardbound volume was first published in 1978 and is still in print. Write for current price and postage from Nimbus Publishing, Limited, 3731 Mackintosh Street, Halifax, Nova Scotia, Canada.

COTSWOLD PRIVIES
by Mollie Harris
When Mollie Harris was a child in the Oxfordshire countryside in the 1920's, the family lavatory was an earth closet at the end of the garden, emptied from time to time by the man with the 'lavender cart'. Few of these ancient lavatories have survived into the 1980's, so Mollie and her collaborator, Sue Chapman, decided to record a dying species before it was too late. Vivid, anecdotal, gruesome, crammed with outlandish information, shamingly funny, *Cotswold Privies* is a descriptive and pictorial tribute to the lavatories of yesteryear, taking us from the unpleasing middle Ages ('Beware of emptynge pysse pottes, and pyssing in chymnes') to the misadventures of P.C. Pike who reported on duty after dropping his helmet down the old vault privy ('By God, Pike, have you stepped in something? There's a most terrible smell in here'.) This charming 80 page paperback was originally published in 1984 by Chatto and Windus-Hogarth Press of England, and in the U.S. by Salem House, 462 Boston Street, Topsfield, MA 01983.

CHAMBERS OF DELIGHT
by Lucinda Lambton
Whoever said the English are prudes hasn't read much of their recent fiction! Lucy's first book, a bestseller, was *Temples of Convenience,* a lavish photo history of the necessary room. *Chambers of Delight,* a fifty-six page volume, is devoted exclusively to chamber pots (thunder mugs), of every hue and "off-color". More facts than you would ever hope to find on a somewhat stale subject and quite lavishly illustrated by Ms. Lambton's exquisite photography. Published in 1983 by The Gordon Fraser Gallery Ltd. London, England.

THE PENNSYLVANIA GERMAN FAMILY FARM
by Amos Long, Jr.

"Der Scheisshaus" was the most important building besides the cow barn on early Pennsylvania farms. The author of this classic 1972 Study devotes a full fifteen pages in volume six to a lively dissertation on privy construction, folklore and first-person anecdotes. Mr. Long documents all of the household utensils which were stored in outdoor necessary rooms and describes many uses (other than the obvious ones) to which this little building was put. The Breinigsville, Pennsylvania German Society.

BACKHOUSES OF THE NORTH
by Muriel E. Newton-White

Thirty-eight pages of lighthearted humor, philosophy and dooley drawings. It effectively captures the spirit of the good old days. To order directly from the publisher send four dollars to Highway Book Shop, Cobalt, Ont., Canada POJ 1 CO.

JOHNS, THE OUTHOUSE BEAUTIFUL
by Frank O'Beirne

Apparently author/illustrator O'Beirne and his publisher, Louis Mariano, of Chicago, ran out of anything to do in the spring of 1952. Why else would they spend a small fortune on a nonsensical book dedicated to full page, super-slick drawings of futuristic, space age outhouses? You can find a copy at most any used book store.

SITTIN' AND A-THINKIN'
by Ernst Peterson and Glenn Chaffin

First published in February of 1953, this classic seventy-two page paperback had gone through its seventh printing by January of 1960. Clever captions and great photographs of "such things as are worth preserving" make it a sought after addition to any privy collector's library. The Dietz Press, Richmond, Virginia.

THE COUNTRY PLUMBER
by Phil Potts

Another 1930's era booklet attempting to cash in on Charles Sale's success. This twenty-eight page volume is nicely done with original woodcuts by an anonymous artist of considerable talent. The story line is about Phil Potts' experiences as a master privy builder. Fly leaf says "copyright applied for" by The Country Press, Inc. Author, publisher and Distributor, Minneapolis, Miss. (no date)

THE SMALLEST ROOM
by John Pudney

If you have always wondered how they got rid of "it" on early trains, planes, boats, space ships and at royal coronations; this 150 page hardbound book is for you. Written from a decidedly British slant, it goes as far back in history as there are any references to public conveniences. Essential reading for all privy trivia buffs. First printed in the United States in 1954 by Hastings House Publishers, New York.

THE HISTORY OF SANITATION
by Bridgeport Brass Co.

"Plumbing Then and Now" is the subtitle of this beautifully illustrated forty-two page paperback published by the Bridgeport Brass Company in 1927. It covers everything from ancient Egyptian temple drains, to Roman baths and 2000 B.C. flush toilets on the island of Crete. Obviously a very scarce piece of advertising ephemera. Try an antiquarian bookseller for this one.

MUDDLED MEANDERINGS IN AN OUTHOUSE
by Bob Ross

Two inexpensive seventy-page volumes of reflections, poems and photos by Mr. Ross and his friends along with colorful cover art by Stan Lynde, creator of the "Rick O'Shay" comic strip. Third printing was in August of 1975. At that time copies were available directly from Bob Ross, 306 East Story, Bozeman, MT 59715.

THE SPECIALIST
by Charles Sale

Who ever thought that this little thirty-page booklet would become a best seller within two months of its original publication in 1929? It was an instant hit and has been continuously in print ever since, having sold over two and a half million copies to date. It is the comical story of an Illinois privy builder, Lem Putt, and how he went about his specialized trade. The book is still available from the Specialist Publishing Co., 109 La Mesa Drive, Burlingame, CA 94010, for five dollars, postpaid.

I'LL TELL YOU WHY
by Charles Sale

A charming little thirty-five page sequel to Chic's first book, *The Specialist,* with drawings by Percy Vogt. Includes halloween pranks and the building of a two-story privy, along with an account of the emerging competition from Bart Wheeler, a local post-hole digger. First published in 1930 and still available for five dollars from the Specialist Publishing Co., 109 La Mesa Drive, Burlingame, CA 94010

UNCLE SAM GOES SPECIALIST
by W. O. Saunders

A spoof on the WPA privy-building boondoggle, published in 1935 by the Elizabeth City Independent newspaper of North Carolina. Thirty-two pages of poems, drawings and commentary, plus copies of U.S. Public Health Service reports. An interesting contemporary overview of the outhouse business.

HOLD EVERYTHING
by Bob Sherwood

This little fifteen-page volume was published late in 1929 in hopes, we assume, of riding the crest of Chic Sale's best seller of the same genre. Mr. Sherwood, a retired clown from P.T. Barnum's circus, was seventy years old at the time.
In his forward he claims to have been a drinking buddy of James Whitcomb Riley during the 1890's and takes credit for being the first to present the hoosier poet's previously suppressed work, *The Old Backhouse,* in copyrighted form. Pen and ink illustrations of comical outhouse situations make this very scarce hardcover booklet worth seeking out from an antiquarian book seller. Originally published in 1929 by Shorewood's, New York City.

THE BATHROOM
by Alexander Kira

Not a book you would want to leave lying around for kids or casual company to browse through; this 271-page bible of bathroom engineering specifications is packed full of explicit photos and uncensored historical prose — all in the name of making modern bathrooms more comfortable, functional and sanitary. For architects, designers, and voyeristic laymen. Copyright 1976, 2nd Edition. Viking Press Inc. New York, NY

PRIVY: THE CLASSIC OUTHOUSE BOOK
by Janet & Richard Strombeck

Originally compiled to sell full-scale floor plans for contemporary privy-style outbuildings; this delightful 8½ by 11 inch volume has been expanded to a full ninety-two pages. Complete with eighty professional renderings including twenty-five floor plans. We found our copy in a dentist's office, but you can order yours for $7.95 plus $1.50 postage from Sterling Publishing Co., Inc., 2 Park Avenue, New York, NY 10016.

VICTORIAN PLUMBING FIXTURES
For Bathrooms and Kitchens

A beautifully reproduced antique trade catalog for artists, designers and researchers. In 1888 the J. L. Mott Iron Works was unquestionably a leader in the bathroom business and over one hundred forty-four water closets are illustrated in this large 277 page paperback. At Bookstores for $15.95 or directly from Dover Publications, Inc., 31 East Second Street, Mineola, NY 11501.

OLD FRIENDS IS ALWAYS BEST
by Paul Webb

An original outhouse saga set in Hillbilly Hollow, U.S.A. Illustrated by the same artist who drew those long-bearded country folk in early Esquire Magazine cartoons. Brown & Bigelow produced this thirty-two page advertising "give away" pamphlet in 1945. Today a worn out, dog-eared copy will set you back at least fifteen dollars.

OUTHOUSE HUMOR
by Billy Edd Wheeler

A new paperback just published by August House, the folklore people from Little Rock, Arkansas. Jokes, stories, songs, and poems about outhouses and thunder mugs, corncobs and honey-dippers, wasps and spiders, and of course, the Sears & Roebuck catalog. All collected over the years by songwriter Billy Edd Wheeler. Copyright 1988. $5.95 at bookstores.

PRETTY PRIVIES OF THE OZARKS
by Mahlon N. White

Twenty-two pages of charming outhouse photos, taken in the 1950's. Still available from the Democrat Publishing Co. 216 South Washington, Clinton, Missouri 64735. (Three dollars including postage).

OF POTS AND PRIVIES
by Makin Wynn (nom de plume)

The lively autobiography of an inquisitive Georgia youth on a trip to Europe. Midway across the continent he became bored with the programmed tour and took a room in Frau Schmidt's rundown hotel. There he meets the village plumber, a privy-lore collector. Together they eat, drink, and spend all their waking hours discussing the history of pots, privies and flushers. Eighty pages of facts and fiction you will never forget. Copyrighted in 1959 by William W. Denlinger of Middleburg, Virginia. Printed in Canada by General Publishing of Toronto.

THE TWO-STORY OUTHOUSE
by Norm Weis

Ghost town wanderings over twelve states and four Canadian provinces fill this recently published 270 page volume. Norm claims it was all done in a twelve year quest for double-decker outhouses, but we detected a lot of authentic mining camp history mixed in with the fascinating privy lore. A delightful book, filled with colorful anecdotes and great black and white photographs. Copyright 1988, The Caxton Printers Ltd. P.O. Box 700, Caldwell, Idaho 83606 ($14.50 postpaid).

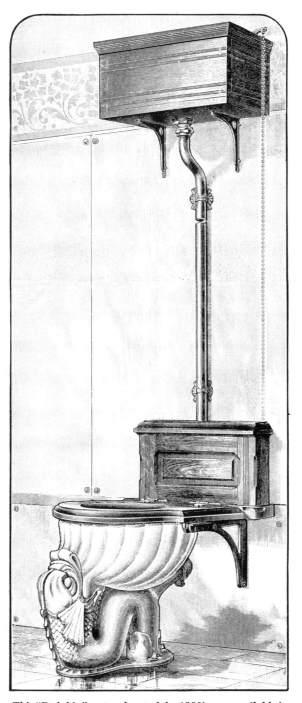

This "Dolphin" water closet of the 1880's was available in ivory, white, or hand painted porcelain. Its automatic flushing mechanism was seat-activated; an idea first patented by Thomas Crapper.

R E P R I N T   S E C T I O N

Home Sanitation and Hygiene (1908) 99

Farm Sewerage Bulletin (1922) 106

National Park Privy Details (1939) 127

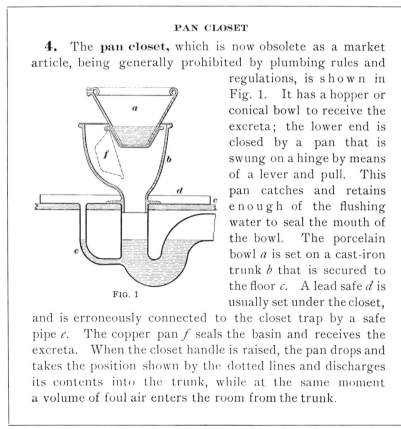

PAN CLOSET

4. The **pan closet,** which is now obsolete as a market article, being generally prohibited by plumbing rules and regulations, is s h o w n in Fig. 1.   It has a hopper or conical bowl to receive the excreta; the lower end is closed by a pan that is swung on a hinge by means of a lever and pull.   This pan catches and retains e n o u g h of the flushing water to seal the mouth of the bowl.   The porcelain bowl a is set on a cast-iron trunk b that is secured to the floor c.   A lead safe d is usually set under the closet, and is erroneously connected to the closet trap by a safe pipe e.   The copper pan f seals the basin and receives the excreta.   When the closet handle is raised, the pan drops and takes the position shown by the dotted lines and discharges its contents into the trunk, while at the same moment a volume of foul air enters the room from the trunk.

FIG. 1

1908

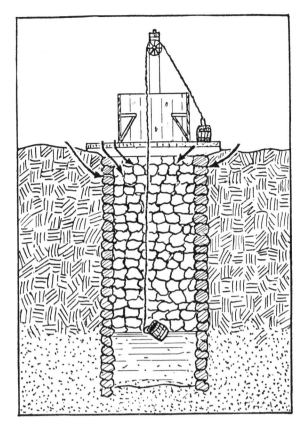

A Typical Insanitary Shallow Well. Filth enters such a well through cracks in the platform, is washed into it by surface water through holes under the platform, seeps into it through the loose casing, and is carried in by the bucket or the rope soiled by filthy hands.

Where water sewage is not available, proper sanitation demands a strictly separate method of disposal of the three kinds of waste matters, namely, solid excreta, slops and garbage. The garbage should be fed to pigs or chicken, burned or buried in a trench at a suitable distance from dwellings and sources of water supply. Kitchen drainage and domestic slops should invariably be disposed of by one of the methods above recommended. They should never be thrown on the surface of the ground near the house or well, allowed to accumulate in an open drain or pool, or deposited in an open privy vault. The cost of a simple system of disposal of these liquid wastes is slight and the danger to health requires that this, at least, be done at any sacrifice.

The Sanitary Privy.—A recent investigation of the subject of soil pollution from open vaults of the ordinary type has been made by Charles Wardell Stiles in connection with his study of the spread of the hookworm disease, especially in the southern states. The following suggestions are condensed and adapted from his report to the Surgeon-General. This plan if generally adopted throughout the United States would eliminate a nuisance which is practically universal and which is perhaps the greatest menace to health now existing in the vicinity of most rural dwellings.

HOW TO BUILD A PRIVY

The following are the essential features: There is a closed portion (box) under the seat for the reception (in a receptacle) and safeguard-

The average style of privy found in the South. It is known as a surface privy, open in back. Notice how the soil pollution is being spread, and how flies can carry the filth to the house and thus infect the food.

ing of the excreta; a room for the occupant; and, proper ventilation.

The receptacle consists practically of a box, with a top represented by the seat, with a floor which is a continuation of the floor of the room, with a front extending from the seat to the floor, with a hinged back which should close tightly, and with two sides continuous with the sides of the room and provided with wire screened ventilators, the upper margin of which is just under the level of the seat. The seat should have one or more holes according to the size of the privy desired, and each hole should have a hinged lid which lifts up toward the back of the room; there should be a piece of wood nailed across the back, on the inside of the room, so as to prevent the lids from being lifted sufficiently to fall backward and so as to make them fall forward of their own accord as soon as the person rises. In this box there should be one or more water-tight tubs, half barrels, pails, or galvanized cans, corresponding to the number of holes in the seat. This receptacle should be high enough to reach nearly to the seat, or, better still, so as to fit snugly against the seat, in order

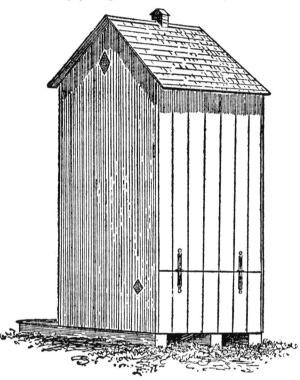

A sanitary privy showing firmly closed door, thus preventing flies, animals, etc., from having access to the fecal material.

to protect the floor against soiling, and sufficiently deep to prevent splashing the person on the seat; it should be held in place by cleats nailed to the floor in such a way that the tub will always be properly centered. The back should be kept closed, as shown in the illustration.

The room should be water-tight and should be provided in front with a good, tightly fitting door. The darker this room can be made the fewer flies will enter. The roof may have a single slant, or a double slant, but while the double slant is somewhat more sightly, the single slant is less expensive in first cost. The room should be provided with two or three wire-screened ventilators, as near the roof as possible.

The ventilators are very important additions to the privy, as they permit a free circulation of air and thus not only reduce the odor but make the outhouse cooler. These ventilators should be copper wire screened in order to keep out flies and other insects. There should be at least 4 (better 5) ventilators, arranged as follows: One each side of the box; one each side the room near the roof; and a fifth ventilator, over the door, in front, is advisable.

Latticework, Flowers and Vines.— At best, the privy is not an attractive addition to the yard. It is possible, however, to reduce its unattractiveness by surrounding it with a latticework on which are trained vines or flowers. This plan, which adds but little to the expense, renders the building much less unsightly and much more private.

Disinfectant.—It is only in comparatively recent years that the privy has been thought worthy of scientific study, and not unnaturally there is some difference of opinion at present as to the best plan to follow in regard to disinfectants.

Top Soil.—Some persons prefer to keep a box or a barrel of top soil, sand, or ashes in the room and to recommend that each time the privy is used the excreta be covered with a shovelful of the dirt. While this has the advantage of simplicity, it has the disadvantage of favoring carelessness, as people so commonly (in fact, as a rule) fail to cover the excreta; further, in order to have the best results, it is necessary to cover the discharges very completely; finally, at best, our knowledge as to how long certain germs and spores will live under these conditions is very unsatisfactory.

Lime. — Some persons prefer to have a box of lime in the room and to cover the excreta with this material. Against this system there is the objection that the lime is not used with sufficient frequency or liberality to keep insects away, as is shown by the fact that flies carry the lime to the house and deposit it on the food.

Water and Oil.—A very cheap and simple method is to pour into the tub about 2 or 3 inches of water; this plan gives the excreta a chance to ferment and liquefy so that the disease germs may be more easily destroyed. If this plan is followed a cup of oil (kerosene will answer) should be poured on the water in order to repel insects.

Cresol.—Some persons favor the use of a 5 per cent crude carbolic acid in the tub, but probably the compound solution of cresol (U. S. P.) will be found equally or more satisfactory if used in a strength of 1 part of this solution to 19 parts of water.

If a disinfectant is used the family should be warned to keep the reserve supply in a place that is not accessible to the children, otherwise accidents may result.

Cleaning the Receptacle.—The frequency of cleaning the receptacle depends upon (a) the size of the tub, (b) the number of persons using the privy, and (c) the weather. In general, it is best to clean it about once a week in winter and twice a week in summer.

An excellent plan is to have a double set of pails or tubs for each privy. Suppose the outhouse is to be cleaned every Saturday: Then pail No. 1 is taken out (say January 1), covered, and set aside until the following Saturday; pail No. 2 is placed in the box for use; on January 8 pail No. 1 is emptied and put back in the box for use while pail No. 2 is taken out, covered, and set aside for a week (namely, until January 15); and so on throughout the year. The object of this plan is to give an extra long time for the germs to be killed by fermentation or by the action of the disinfectant before the pail is emptied.

Each time that the receptacle is emptied, it is best to sprinkle into it a layer of top soil about a quarter to half an inch deep before putting it back into the box.

Disposal of the Excreta.—For the present, until certain very thorough investigations are made in regard to the length of time that the eggs of parasites and the spores of certain other germs may live, it is undoubtedly best to burn or boil all excreta; where this is not feasible, it is best to bury all human discharges at least 300 feet away and down hill from any water supply (as the well, spring, etc.).

Many farmers insist upon using the fresh night soil as fertilizer. In warm climates this is attended with considerable danger, and if it is so utilized, it should never be used upon any field upon which vegetables are grown which are eaten uncooked; further, it should be promptly plowed under.

In our present lack of knowledge as to the length of time that various germs (as spores of the ameba which produce dysentery, various eggs, etc.) *may live, the use of fresh, unboiled night soil as a fertilizer is false economy which may result in loss of human life. This is especially true in warm climates.*

Directions for Building a Sanitary Privy.—In order to put the construction of a sanitary privy for the home within the carpentering abilities of boys, a practical carpenter has been requested to construct models to conform to the general ideas expressed in this article, and to furnish estimates of the amount of lumber, hardware, and wire screening required. Drawings of these models have been made during the process of construction and in completed condition. The carpenter was requested to hold constantly in mind two points, namely, economy and simplicity of construction. It is believed that any 14-year-old schoolboy of average intelligence and mechanical ingenuity can, by following these plans, build a sanitary privy for his home at an expense for building materials, exclusive of receptacle, of $5 to $10, according to locality. It is further believed that the plans submitted cover the essential points to be considered. They can be elaborated to suit the individual taste of persons who prefer a more elegant and more expensive structure. For instance, the roof can have a double

The sanitary privy. Front view.

instead of a single slant, and can be shingled; the sides, front, and back can be clapboarded or they can be shingled. Instead of one seat, there may be two, three, four, or five seats, etc., according to necessity.

A Single-Seated Privy for the Home.—Nearly all privies for the home have seats for two persons, but a single privy can be made more economically.

Framework.—The lumber required for the framework of the outhouse shown is as follows:

A. Two pieces of lumber (scantling) 4 feet long and 6 inches square at ends.

B. One piece of lumber (scantling) 3 feet 10 inches long; 4 inches square at ends.

C. Two pieces of lumber (scantling) 3 feet 4 inches long; 4 inches square at ends.

D. Two pieces of lumber (scantling) 7 feet 9 inches long; 2 by 4 inches at ends.

E. Two pieces of lumber (scantling) 6 feet 7 inches long; 2 by 4 inches at ends.

F. Two pieces of lumber (scantling) 6 feet 3 inches long; 2 by 4 inches at ends.

G. Two pieces of lumber (scantling) 5 feet long; 2 by 4 inches at ends.

H. One piece of lumber (scantling) 3 feet 10 inches long; 2 by 4 inches at ends.

I. Two pieces of lumber (scantling) 3 feet 4 inches long; 2 by 4 inches at ends.

J. Two pieces of lumber (scantling) 3 inches long; 2 by 4 inches at ends.

K. Two pieces of lumber (scantling) 4 feet 7 inches long; 6 inches wide by 1 inch thick. The ends of K should be trimmed after being nailed in place.

L. Two pieces of lumber (scantling) 4 feet long, 6 inches wide, and 1 inch thick.

First lay down the sills marked A and join them with the joist marked B; then nail in position the two joists marked C, with their ends 3 inches from the outer edge of A; raise the corner posts (D and F), spiking them at bottom to A and C, and joining them with L, I_2, G, and K; raise door posts E, fastening them at J, and then spike I_1 in position; H is fastened to K.

Sides. — Each side requires four boards (a) 12 inches wide by 1 inch thick and 8 feet 6 inches long; these are nailed to K, L, and A. The corner boards must be notched at G, allowing them to pass to bottom of roof; next draw a slant from front to back at G-G, on the outside of the boards, and saw the four side boards to correspond with this slant.

Back. — The back requires two boards (b) 12 inches wide by 1 inch thick and 6 feet 11 inches long, and two boards (c) 12 inches wide by 1 inch thick and 6 feet 5 inches long. The two longest boards (b) are nailed next to the sides; the shorter

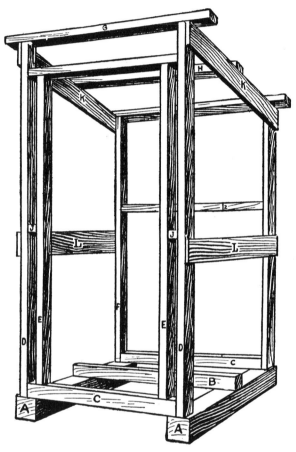

Framework of the sanitary privy.

boards (c) are sawed in two so that one piece (c1) measures 4 feet 6 inches, the other (c2) 1 foot 11 inches; the longer portion (c1) is nailed in position above the seat; the shorter portion (c2) is later utilized in making the back door.

Floor. — The floor requires four boards (d) which (when cut to fit) measure 1 inch thick, 12 inches wide, and 3 feet 10 inches long.

Front.—The front boards may next be nailed on. The front requires (aside from the door) two boards (e) which (when cut to fit) measure

1 inch thick, 9 inches wide, and 8 feet 5 inches long; these are nailed next to the sides.

Roof.—The roof may now be finished. This requires five boards (f) measuring (when cut to fit) 1 inch thick, 12 inches wide, and 6 feet long. They are so placed that they extend 8 inches beyond the front. The joints (cracks) are to be broken (covered) by laths one-half inch thick, 3 inches broad, and 6 feet long.

Box.—The front of the box requires two boards, 1 inch thick and 3 feet 10 inches long. One of these (g) may measure 12 inches wide, the other (h) 5 inches wide. These are nailed in place, so that the back of the boards is 18 inches from the inside of the backboards. The seat of the box requires two boards, 1 inch thick, 3 feet 10 inches long; one of these (i) may measure 12 inches wide, the other (j) 7 inches wide. One must be jogged (cut out) to fit around the back corner posts (F). An oblong hole, 10 inches long and $7\frac{1}{2}$ inches wide, is cut in the seat. The edge should be smoothly rounded or beveled. An extra (removable) seat for children may be made by cutting a board 1 inch thick, 15 inches wide, and 20 inches long; in this seat a hole is cut, measuring 7 inches long by 6 inches wide; the front margin of this hole should be about 3 inches from the front edge of the board; to prevent warping, a cross cleat is nailed on top near or at each end of the board.

A cover (k) to the seat should measure 1 inch thick by 15 inches wide by 20 inches long; it is cleated on top near the ends, to prevent warping; it is hinged in back to a strip 1 inch thick, 3 inches wide, and 20 inches long, which is fastened to the seat. Cleats (m) may also be nailed on the seat at the sides of the cover. On the inside of the backboard, 12 inches above the seat, there should be nailed a block (l), 2 inches thick, 6 inches long, extending forward $3\frac{1}{4}$ inches; this is intended to prevent the cover from falling backward and to make it to fall down over the hole when the occupant rises.

On the floor of the box (underneath the seat) two or three cleats (n) are nailed in such a position that they will always center the tub; the position of these cleats depends upon the size of the tub.

Back Door.—In making the back of the privy the two center boards (c) were sawed at the height of the bottom of the seat. The small portions (c²) sawed off (23 inches long) are cleated (o) together so as to

The sanitary privy. Rear and side view.

form a back door which is hinged above; a bolt or a button is arranged to keep the door closed.

Front Door.—The front door is made by cleating (p) together three boards (q) 1 inch thick, 10 inches wide, and (when finished) 6 feet 7 inches long; it is best to use three cross cleats (p) (1 inch thick, 6 inches wide, 30 inches long), which are placed on the inside. The door is hung with two hinges (6-inch "strap" hinges will do), which are placed on the right as one faces the privy, so that the door opens from the left. The door should close with a coil spring (cost about 10 cents)

or with a rope and weight, and may fasten on the inside with a catch or a cord. Under the door a cross-piece (r) 1 inch thick, 4 inches wide, 30 inches long (when finished) may be nailed to the joist. Stops (s) may be placed inside the door as illustrated in the cut. These should be 1 inch thick, 3 inches wide, and 6 feet 6 inches long, and should be jogged (cut out) (t) to fit the cross cleats (p) on the door. Close over the top of the door place a strip (v) 1 inch thick, 2 inches wide, 30 inches long, nailed to I. A corresponding piece (v) is placed higher up directly under the roof, nailed to G. A strap or door pull is fastened to the outside of the door.

Ventilators.—There should be five ventilators (w). One is placed at each side of the box, directly under the seat; it measures 6 to 8 inches square. Another (12 inches square) is placed near the top on each side of the privy. A fifth (30 inches long, 8½ inches wide) is placed over the door, between G and I_1. The ventilators are made of 15-mesh copper wire, which is first tacked in place and then protected at the edge with the same kind of lath that is used on the cracks and joints.

Lath.—Outside cracks (joints) are covered with lath one-half inch thick by 3 inches wide.

Receptacle.—For a receptacle, saw a water-tight barrel to fit snugly under the seat; or purchase a can or tub, as deep (17 inches) as the distance from the under surface of the seat to the floor. If it is not possible to obtain a tub, barrel, or can of the desired size, the receptacle used should be elevated from the floor by blocks or boards so that it fits snugly under the seat. A galvanized can measuring 16 inches deep and 16 inches in diameter can be purchased for about $1, or even less. An empty candy bucket can be purchased for about 10 cents.

Order for Material.—The carpenter has made out the following order for lumber (pine, No. 1 grade) and hardware to be used in building a privy such as here illustrated:

1 piece scantling, 6 by 6 inches by 8 feet long, 24 square feet.

1 piece scantling, 4 by 4 inches by 12 feet long, 16 square feet.

5 pieces scantling, 2 by 4 inches by 16 feet long, 54 square feet.

3 pieces board, 1 by 6 inches by 16 feet long, 24 square feet.

2 pieces board, 1 by 9 inches by 9 feet long, 14 square feet.

3 pieces board, 1 by 10 inches by 7 feet long, 18 square feet.

15 pieces board, 1 by 12 inches by 12 feet long, 180 square feet.

12 pieces board, ½ by 3 inches by 16 feet long, 48 square feet.

2 pounds of 20-penny spikes.

6 pounds of 10-penny nails.

2 pounds of 6-penny nails.

7 feet screen, 15-mesh, copper, 12 inches wide.

4 hinges, 6-inch "strap," for front and back doors.

2 hinges, 6-inch "T," or 3-inch "butts," for cover.

1 coil spring for front door.

According to the carpenter's estimate, these materials will cost from $5 to $10, according to locality.

There is some variation in the size of lumber, as the pieces are not absolutely uniform. The sizes given in the lumber order represent the standard sizes which should be ordered, but the purchaser need not expect to find that the pieces delivered correspond with mathematical exactness to the sizes called for. On this account the pieces must be measured and cut to measure as they are put together.

Elimination of Flies.—A link between the subject of home sanitation and hygiene and that of the prevention of disease has been forged by the discovery that the deadly germs of enteric diseases,—such as typhoid fever, cholera, cholera infantum and tropical dysentery—are frequently communicated to man by the common house fly. Other diseases which are less commonly transmitted by flies are tuberculosis, anthrax, bu-

bonic plague (black death), trachoma, septicemia, erysipelas, leprosy, yaws, and, perhaps, smallpox. The problem of eliminating the house fly belongs to the subject of sanitation because flies commonly become infected with noxious bacteria from feeding upon infected garbage or

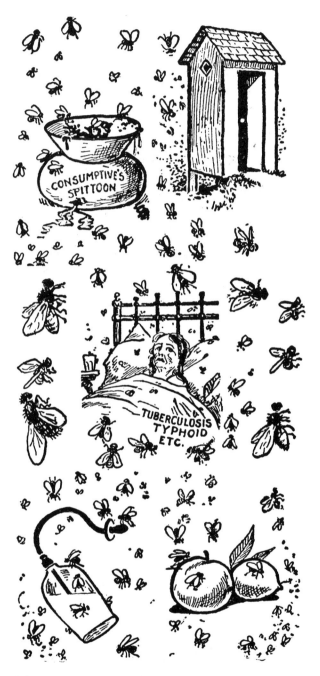

The agency of flies in communicating disease. Courtesy of the State Board of Health of Florida.

other domestic refuse or the excreta of persons suffering from typhoid or other communicable disease, or those of healthy carriers. The elimination of these nuisances by the various methods of disposal above recommended is half the battle in the prevention of disease.

U. S. DEPARTMENT OF AGRICULTURE

FARMERS' BULLETIN No. 1227

SEWAGE AND SEWERAGE OF FARM HOMES

GEORGE M. WARREN,

Engineer, Bureau of Public Roads

Issued January, 1922
Revised October, 1928

	Page		Page
Introduction	1	Importance of air in treatment of sewage	7
Sewage, sewers, and sewerage defined	1	Practical utilities	8
Nature and quantity of sewage	2	Septic tanks	21
Sewage-borne diseases and their avoidance	2	Grease traps	43
How sewage decomposes	5	General procedure	45

INTRODUCTION

The main purpose of home sewerage works is to get rid of sewage in such way as (1) to guard against the transmission of disease germs through drinking water, flies, or other means; (2) to avoid creating nuisance. What is the best method and what the best outfit are questions not to be answered offhand from afar. A treatment that is a success in one location may be a failure in another. In every instance decision should be based upon field data and full knowledge of the local needs and conditions. An installation planned from assumed conditions may work harm. The householder may be misled as to the purification and rely on a protection that is not real. He may anticipate little or no odor and find a nuisance has been created.

SEWAGE, SEWERS, AND SEWERAGE DEFINED

Human excrements (feces and urine) as found in closets and privy vaults are known as night soil. These wastes may be flushed away with running water, and there may be added the discharges from washbasins, bathtubs, kitchen and slop sinks, laundry trays, washing vats, and floor drains. This refuse liquid product is sewage, and the underground pipe which conveys it is a sewer. Since sewers carry foul matter they should be water-tight, and this feature of their construction distinguishes them from drains removing relatively pure surface or ground water. Sewerage refers to a system of sewers, including the pipes, tanks, disposal works, and appurtenances.

Under average conditions a man discharges daily about 3½ ounces of moist feces and 40 ounces of urine, the total in a year approximating 992 pounds.[1] Feces consist largely of water and undigested or partially digested food; by weight it is 77.2 per cent water.[2] Urine is about 96.3 per cent water.[2]

The excrements constitute but a small part of ordinary sewage. In addition to the excrements and the daily water consumption of perhaps 40 gallons per person are many substances entering into the economy of the household, such as grease, fats, milk, bits of food, meat, fruit and vegetables, tea and coffee grounds, paper, etc. This complex product contains mineral, vegetable, and animal substances, both dissolved and undissolved. It contains dead organic matter and living organisms in the form of exceedingly minute vegetative cells (bacteria) and animal cells (protozoa). These low forms of life are the active agents in destroying dead organic matter.

The bacteria are numbered in billions and include many species, some useful and others harmful. They may be termed tiny scavengers, which under favorable conditions multiply with great rapidity, their useful work being the oxidizing and nitrifying of dissolved organic matter and the breaking down of complex organic solids to liquids and gases. Among the myriads of bacteria are many of a virulent nature. These at any time may include species which are the cause of well-known infectious and parasitic diseases.

SEWAGE-BORNE DISEASES AND THEIR AVOIDANCE

Any spittoon, slop pail, sink drain, urinal, privy, cesspool, sewage tank, or sewage distribution field is a potential danger. A bit of spit, urine, or feces the size of a pin head may contain many hundred germs, all invisible to the naked eye and each one capable of producing disease. These discharges should be kept away from the food and drink of man and animals. From specific germs that may be carried in sewage at any time there may result typhoid fever, tuberculosis, cholera, dysentery, diarrhea, and other dangerous ailments, and it is probable that other maladies may be traced to human waste. From certain animal parasites or their eggs that may be carried in sewage there may result intestinal worms, of which the more common are the hookworm, roundworm, whipworm, eelworm, tapeworm, and seat worm.

Sewage, drainage, or other impure water may contain also the causative agents of numerous ailments common to livestock, such as tuberculosis, foot-and-mouth disease, hog cholera, anthrax, glanders, and stomach and intestinal worms.

Disease germs are carried by many agencies and unsuspectingly received by devious routes into the human body. Infection may come from the swirling dust of the railway roadbed, from contact with transitory or chronic carriers of disease, from green truck grown in gardens fertilized with night soil or sewage, from food prepared or touched by unclean hands or visited by flies or vermin,

---

[1] Practical Physiological Chemistry, by Philip B. Hawk, 1916, pp. 221, 359.
[2] Agriculture, by F. H. Storer, 1894, vol. 2, p. 70.

from milk handled by sick or careless dairymen, from milk cans and utensils washed with contaminated water, or from cisterns, wells, springs, reservoirs, irrigation ditches, brooks, or lakes receiving the surface wash or the underground drainage from sewage-polluted soil.

Many recorded examples show with certainty how typhoid fever and other diseases have been transmitted. A few indicating the responsibilities and duties of people who live in the country are cited here.

In August, 1889, a sister and two brothers aged 18, 21, and 23 years, respectively, and all apparently in robust health dwelt together in a rural village in Columbiana County, Ohio. Typhoid fever in particular virulent form developed after use of drinking water from a badly polluted surface source. The deaths of all three occurred within a space of 10 days.

In September and October, 1899, 63 cases of typhoid fever, resulting in 5 deaths, occurred at the Northampton (Mass.) insane hospital. This epidemic was conclusively traced to celery, which was eaten freely in August and was grown and banked in a plot that had been fertilized in the late winter or early spring with the solid residue and scrapings from a sewage filter bed situated on the hospital grounds.

Some years ago Dr. W. W. Skinner, Bureau of Chemistry, Department of Agriculture, investigated the cause of an outbreak of typhoid fever in southwest Virginia. A small stream meandered through a narrow valley in which five 10-inch wells about 450 feet deep had been drilled in limestone formation. The wells were from 50 to 400 feet from the stream, from which, it was suspected, pollution was reaching the wells. In a pool in the stream bed approximately one-fourth mile above the wells several hundred pounds of common salt were dissolved. Four of the wells were cut off from the pump and the fifth was subjected to heavy pumping. The water discharged by the pump was examined at 15-minute intervals and its salt content determined over a considerable period of time. After the lapse of several 15-minute intervals the salt began to rise and continued to rise until the maximum was approximately seven times that at the beginning of the test, thus proving the facility with which pollution may pass a long distance underground and reach deep wells.

Probably no epidemic in American history better illustrates the dire results that may follow one thoughtless act than the outbreak of typhoid fever at Plymouth, Pa., in 1885. In January and February of that year the night discharges of one typhoid fever patient were thrown out upon the snow near his home. These, carried by spring thaws into the public water supply, caused an epidemic running from April to September. In a total population of about 8,000, 1,104 persons were attacked by the disease and 114 died.

Like plants and animals, disease germs vary in their powers of resistance. Some are hardy, others succumb easily. Outside the body most of them probably die in a few days or weeks. It is never certain when such germs may not lodge where the immediate surroundings are favorable to their life and reproduction. Milk is one of the common substances in which germs multiply rapidly. The experience at Northampton shows that typhoid-fever germs may survive several months in garden soil. Laboratory tests by the United States Public Health Service showed that typhoid-fever germs had not all succumbed after being frozen in cream 74 days. (Public Health Reports, Feb. 8, 1918, pp. 163–166.) Ravenel kept the spores of anthrax immersed for 244 days in the strongest tanning fluids without perceptible change in their vitality or virulence. (Annual Report, State Department of Health, Mass., 1916, p. 494.)

Unsafe practices.—Upon thousands of small farms there are no privies and excretions are deposited carelessly about the premises. A place of this character is shown in figure 1. Upon thousands of other farms the privy is so filthy and neglected that hired men and visitors seek near-by sheds, fields, and woods. A privy of this character is shown in figure 2. These practices and conditions exist in every section of the country. They should be abolished.

Deserving of severe censure is the old custom of conveying excrements or sewage into abandoned wells or some convenient stream. Such a practice is indecent and unsafe. It is unnecessary and is contrary to the laws of most of the States.

Likewise dangerous and even more disgusting is the old custom of using human excrement or sewage for the fertilization of truck land. Under no circumstances should such wastes be used on land devoted to celery, lettuce, radishes, cucumbers, cabbages, tomatoes, melons, or other vegetables, berries, or low-growing fruits that are eaten raw. Disease germs or particles of soil containing such germs may adhere to the skins of vegetables or fruits and infect the eater.

Upon farms it is necessary to dispose of excretal wastes at no great distance from the dwelling. The ability and likelihood of flies

Fig. 1.—One of many farms lacking the simplest sanitary convenience

carrying disease germs direct to the dinner table, kitchen, or pantry are well known. Vermin, household pets, poultry, and live stock may spread such germs. For these reasons, and also on the score of odor, farm sewage never should be exposed.

Important safety measure.—The farmer can do no other one thing so vital to his own and the public health as to make sure of the continued purity of the farm water supply. Investigations indicate that about three out of four shallow wells are polluted badly.

Wells and springs are fed by ground water, which is merely natural drainage. Drainage water usually moves with the slope of the land. It always dissolves part of the mineral, vegetable, and animal matter of the ground over or through which it moves. In this way impurities are carried into the ground water and may reach distant wells or springs.

The great safeguards are clean ground and wide separation of the well from probable channels of impure drainage water. It is not enough that a well or spring is 50 or 150 feet from a source of filth or that it is on higher ground. Given porous ground, a seamy ledge, or long-continued pollution of one plat of land, the zone of contamination is likely to extend long distances, particularly in downhill directions or when the water is low through drought or heavy pumping. Only when the surface of the water in a well or spring is at a higher level at all times than any near-by source of filth is there assurance of safety from impure seepage. Some of the foregoing facts are shown diagrammatically in Figure 3. Figure 4 is typical of those insanitary, poorly drained barnyards that are almost certain to work injury to wells situated in or near them. Accumulations of filth result in objectionable odor and noxious drainage. Figure 5 illustrates poor relative location of privy, cesspool, and well.

FIG. 2.—The rickety, uncomfortable, unspeakably f o u l, dangerous ground privy. Neglected by the owner, shunned by the hired man, avoided by the guest, who, in preference goes to near-by fields or woods, polluter of wells, meeting place of house flies and disease germs, privies of this character abide only because of man's indifference

Sewage or impure drainage water should never be discharged into or upon ground draining toward a well, spring, or other source of water supply. Neither should such wastes be discharged into openings in rock, an abandoned well, nor a hole, cesspool, vault, or tank so located that pollution can escape into water-bearing earth or rock. Whatever the system of sewage disposal, it should be entirely and widely separated from the water supply. Further information on locating and constructing wells is given in Farmers' Bulletin 1448F, Farmstead Water Supply, copies of which may be had upon request to the Division of Publications, Department of Agriculture.

Enough has been said to bring home to the reader these vital points:

1. Never allow the farm sewage or excrements, even in minutest quantity, to reach the food or water of man or livestock.

2. Never expose such wastes so that they can be visited by flies or other carriers of disease germs.

3. Never use such wastes to fertilize or irrigate vegetable gardens.

4. Never discharge or throw such wastes into a stream, pond, or abandoned well, nor into a gutter, ditch, or tile drainage system, which naturally must have outlet in some watercourse.

HOW SEWAGE DECOMPOSES

When a bottle of fresh sewage is kept in a warm room changes occur in the appearance and nature of the liquid. At first it is light in appearance and its odor is slight. It is well supplied with oxygen, since this gas is always found in waters exposed to the atmosphere. In a few hours the solids in the sewage separate mechanically according to their relative weights; sediment collects at the bottom, and a greasy film covers the surface. In a day's time there is an enormous

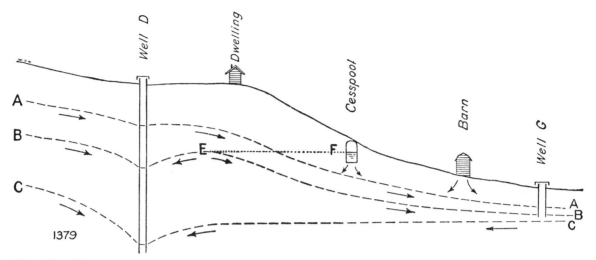

Fig. 3.—How an apparently good well may draw foul drainage. Arrows show direction of ground-water movement. *A–A*, Usual water table (surface of free water in the ground); *B–B*, water table lowered by drought and pumping from well *D*; *C–C*, water table further lowered by drought and heavy pumping; *E–F*, level line from surface of sewage in cesspool. Well *D* is safe until the water table is lowered to *E*; further lowering draws drainage from the cesspool and, with the water table at *C–C*, from the barn. The location of well *G* renders it always unsafe

development of bacteria, which obtain their food supply from the dissolved carbonaceous and nitrogenous matter. As long as free oxygen is present this action is spoken of as aërobic decomposition. There is a gradual increase in the amount of ammonia and a decrease of free

Fig. 4.—An insanitary, poorly drained barnyard. (Board of Health, Milwaukee.) Liquid manure or other foul drainage is sure to leach into wells situated in or near barnyards of this character

oxygen. When the ammonia is near the maximum and the free oxygen is exhausted the sewage is said to be stale. Following exhaustion of the oxygen supply, bacterial life continues profuse, but it gradually diminishes as a result of reduction of its food supply and

111

the poisonous effects of its own wastes. In the absence of oxygen the bacterial action is spoken of as anaërobic decomposition. The sewage turns darker and becomes more offensive. Suspended and settled organic substances break apart or liquefy later, and various foul-smelling gases are liberated. Sewage in this condition is known as septic and the putrefaction that has taken place is called septicization. Most of the odor eventually disappears, and a dark, insoluble, mosslike substance remains as a deposit. Complete reduction of this deposit may require many years.

Fig. 5.—Poor relative locations of privy, cesspool, and well. (State Department of Health, Massachusetts.) Never allow privy, cesspool, or sink drainage to escape into the plot of ground from which the water supply comes

IMPORTANCE OF AIR IN TREATMENT OF SEWAGE

Decomposition of organic matter by bacterial agency is not a complete method of treating sewage, as will be shown later under " Septic tanks." It is sufficient to observe here that in all practical methods of treatment aeration plays a vital part. The air or the sewage, or both, must be in a finely divided state, as when sewage percolates through the interstices of a porous, air-filled soil. The principle involved was clearly stated 30 years ago by Hiram F. Mills, a member of the Massachusetts State Board of Health. In discussing the intermittent filtration of sewage through gravel stones too coarse to arrest even the coarest particles in the sewage Mr. Mills said: " The slow movement of the sewage in thin films over the surface of the stones, with air in contact, caused a removal for some months of 97 per cent of the organic nitrogenous matter, as well as 99 per cent of the bacteria."

Previous discussion has dealt largely with basic principles of sanitation. The construction and operation of simple utilities embodying some of these principles are discussed in the following order: (1) Privies for excrements only; (2) works for handling wastes where a supply of water is available for flushing.

PIT PRIVY

Figure 6 shows a portable pit privy suitable for places of the character of that shown in figure 1, where land is abundant and cheap, and in such localities has proved practical. It provides, at

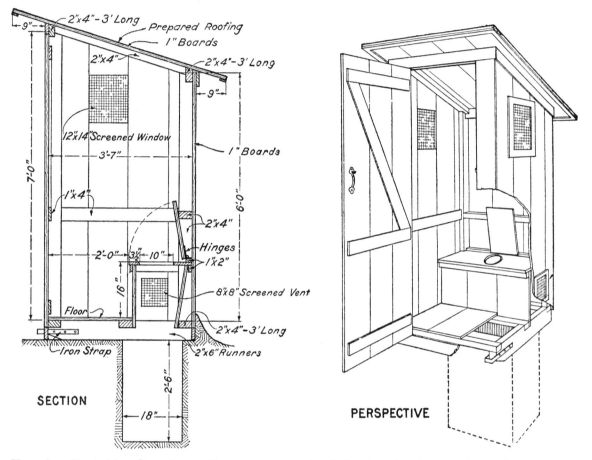

FIG. 6.—Portable pit privy. For use where land is abundant and cheap, but unless handled with judgment can not be regarded as safe. The privy is mounted on runners for convenience in moving to new locations

minimum cost and with least attention, a fixed place for depositing excretions where the filth can not be tracked by man, spread by animals, reached by flies, nor washed by rain.

The privy is light and inexpensive and is placed over a pit in the ground. When the pit becomes one-half or two-thirds full the privy is drawn or carried to a new location. The pit should be shallow, preferably not over 2½ feet in depth, and never should be located in wet ground or rock formation or where the surface or the strata slope toward a well, spring, or other source of domestic water supply. Besides being in lower ground the pit should never be within 200 feet of a well or spring. Since dryness in the pit is essential, the ground should be raised slightly and 10 or 12 inches of earth should be banked and compacted against all sides to shed rain water. The banking also serves to exclude flies. If the soil is sandy or gravelly, the pit should be lined with boards or pales to prevent caving. The privy should be boarded closely and should be provided with screened openings for ventilation and light. The screens may consist of standard galvanized or black enameled wire cloth having 14 squares to the inch. The whole seat should be easily removable for cleaning. A little loose absorbent soil should be added daily to the accumulation in the pit, and when a pit is abandoned it should be filled immediately with dry earth mounded to shed water.

A pit privy for use in field work, consisting of a framework of ½-inch iron pipe for corner posts connected at the top with ¼-inch iron rods bent at the ends to right angles and hung with curtains of unbleached muslin, is described in Public Health Report of the United States Public Health Service, July 26, 1918.

A pit privy, even if moved often, can not be regarded as safe. The danger is that accumulations of waste may overtax the purifying capacity of the soil and the leachings reach wells or springs. Sloping ground is not a guaranty of safety; the great safeguard lies in locating the privy a long distance from the water supply and as far below it as possible.

SANITARY PRIVY

The next step in evolution is the sanitary privy. Its construction must be such that it is practically impossible for filth or germs to be spread above ground, to escape by percolation underground, or to be accessible to flies, vermin, chickens, or animals. Furthermore, it must be cared for in a cleanly manner, else it ceases to be sanitary. To secure these desirable ends sanitarians have devised numerous types of tight-receptacle privy. Considering the small cost and the proved value of some of these types, it is to be regretted that few are seen on American farms.

The container for a sanitary privy may be small—for example, a galvanized-iron pail or garbage can, to be removed from time to time by hand; it may be large, as a barrel or a metal tank mounted for moving; or it may be a stationary underground metal tank or masonry vault. The essential requirement in the receptacle is permanent water-tightness to prevent pollution of soils and wells. Wooden pails or boxes, which warp and leak, should not be used. Where a vault is used it should be shallow to facilitate emptying and cleaning. Moreover, if the receptacle should leak it is better that the escape of liquid should be in the top soil, where air and bacterial life are most abundant.

Sanitary privies are classified according to the method used in treating the excretions, as dry earth, chemical, etc.

DRY-EARTH PRIVY

Pail type.—A very serviceable pail privy is shown in Figures 7 and 8. The method of ventilation is an adaptation of a system that has proved very effective in barns and other buildings here and abroad. A flue with a clear opening of 16 square inches rises from the rear of the seat and terminates above the ridgepole in a cowl or small roofed housing. Attached to this flue is a short auxiliary duct, 4 by 15 inches, for removing foul air from the top of the privy. In its upper portion on the long sides the cowl is open, allowing free movement of air across the top of the flue. In addition, the long sides of the cowl are open below next to the roof. These two openings, with the connecting vertical air passages, permit free upward movement of air through the cowl, as indicated by the arrows. The combined effect is to create draft from beneath the seat and from the top of the privy. The ventilating flue is 2 by 8 inches at the seat and 4 by 4 inches 5 feet above. The taper slightly increases the labor of making the flue, but permits a 2-inch reduction in the length of the building.

In plan the privy is 4 by 4½ feet. The sills are secured to durable posts set about 4 feet in the ground. The boarding is tight, and all vents and windows are screened to exclude insects. The screens may

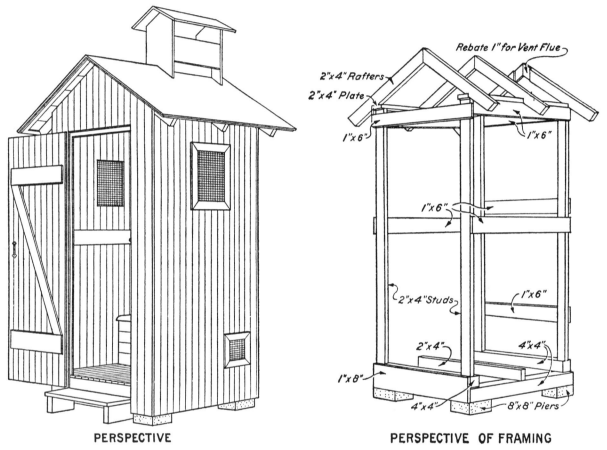

PERSPECTIVE   PERSPECTIVE OF FRAMING

FIG. 7.—Pail privy. Well constructed, ventilated, and screened. With proper care is sanitary and unobjectionable

be the same as for pit privies or, if a more lasting material is desired, bronze or copper screening of 14 squares to the inch may be used. The entire seat is hinged, thus permitting removal of the receptacle and facilitating cleaning and washing the underside of the seat and the destruction of spiders and other insects which thrive in dark, unclean places. The receptacle is a heavy galvanized-iron garbage can. Heavy brown-paper bags for lining the can may be had at slight cost, and their use helps to keep the can clean and facilitates emptying. Painting with black asphaltum serves a similar purpose and protects the can from rust. If the contents are frozen, a little heat releases them. Of nonfreezing mixtures a strong brine made with common salt or calcium chloride is effective. Two and one-half to 3 pounds of either thoroughly dissolved in a gallon of water lowers the freezing point of the mixture to about zero. Denatured alcohol or

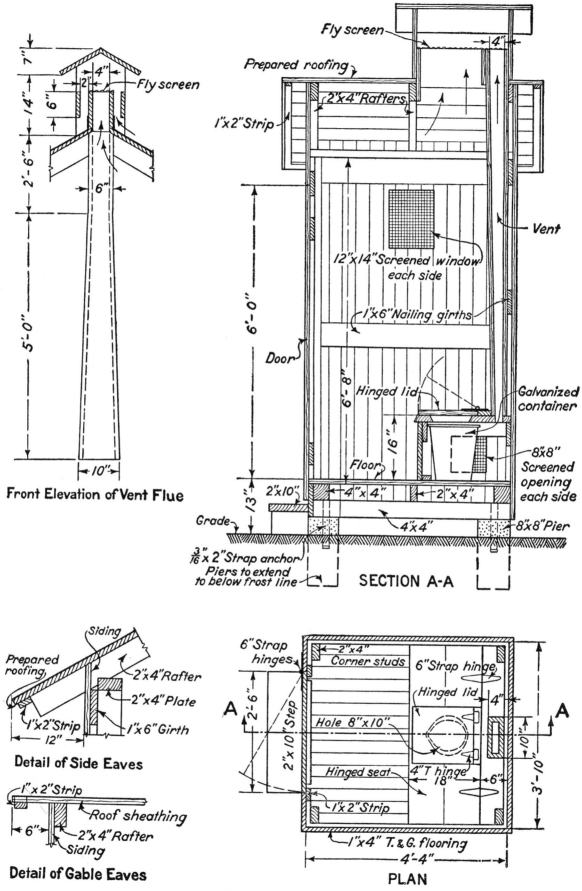

Front Elevation of Vent Flue

7"
4"
2"
6"
1'4"
Fly screen
2'-6"
6"
5'-0"
10"

Fly screen
Prepared roofing
2"x4"Rafters
1"x2"Strip
Vent
12"x14"Screened window each side
1"x6"Nailing girths
Door
6'-0"
Galvanized container
Hinged lid
6'-8"
16"
8"x8" Screened opening each side
Floor
2"x10"
4"x4"
2"x4"
13"
Grade
4"x4"
8"x8"Pier
3/16"x 2"Strap anchor
Piers to extend to below frost line

SECTION A-A

Siding
Prepared roofing
2"x4"Rafter
2"x4"Plate
1"x2"Strip
1"x6"Girth
12"

Detail of Side Eaves

1"x2"Strip
Roof sheathing
6"
2"x4"Rafter
Siding

Detail of Gable Eaves

6"Strap hinges
2"x4"
Corner studs
6"Strap hinge
Hinged lid
4"
A
2'-6"
2"x10"Step
Hole 8"x10"
10"
A
Hinged seat
4"T hinge
18"
6"
3'-10"
1"x2"Strip
1"x4" T.& G. flooring
4'-4"

PLAN

FIG. 8.—Pail privy

116

wood alcohol in a 25 per cent solution has a like low freezing point and the additional merit of being noncorrosive of metals. The can should be emptied frequently and the contents completely buried in a thin layer by a plow or in a shallow hand-dug trench at a point below and remote from wells and springs. Wherever intestinal disease exists the contents of the can should be destroyed by burning or made sterile before burial by boiling or by incorporation with a strong chemical disinfectant.

A privy ventilated in the manner before described is shown in Figure 9. The cowl, however, is open on four sides instead of two sides as shown in Figures 7 and 8. The working drawings (figs. 7 and 8) show that the construction of a privy of the kind is not difficult. Figure 10 gives three suggestions whereby a privy may be conveniently located and the approach screened or partially hidden by latticework, vines, or shrubbery.

FIG. 9.—A well-ventilated privy in Montana

Vault type.—A primitive and yet serviceable three-seat dry-earth privy of the vault type is shown in Figure 11. This privy was constructed in 1817 upon a farm at Westboro, Mass. The vault, made of bricks, was 6 feet long by 5 feet wide, and the bottom was 1 foot below the surface of the ground. The brickwork was laid in mortar, and the part below the ground surface was plastered on the inside. The outside of the vault was exposed to light and air on all four sides. Across the long side of the vault in the rear was a door swinging upward through which the night soil was removed two or three times a year, usually in the spring, summer, and fall, and hauled to a near-by field, where it was deposited in a furrow, just ahead of the plow.

Especial attention is called to the shallowness of the vault and the lightened labor of cleaning it out. The swinging door at the rear facilitated the sprinkling of dry soil or ashes over the contents of the vault, thus avoiding the necessity of carrying dirt and dust into the building and dust settling upon the seat. This privy was in use for nearly 100 years without renewal or repairs. When last seen the original seat, which always was kept painted, showed no signs of decay. Modern methods would call for a concrete vault of guaranteed water-tightness,[3] proper ventilation and screening, and hinging the seat.

Working drawings for a very convenient well-built two-seat vault privy are reproduced in Figures 12 and 13. The essential features

[3] Directions for mixing and placing concrete to secure water-tightness are contained in Farmers' Bulletin 1279–F, "Plain concrete for farm use," and Farmers' Bulletin 1572–F, "Making Cellars Dry."

are shown in sufficient detail to require little explanation. With concrete mixtures of $1:2:3$ (1 volume cement, 2 volumes sand, 3 volumes stone) for the vault and $1:2:4$ for the posts there will be required a total of about 2 cubic yards of concrete, taking 3½ barrels

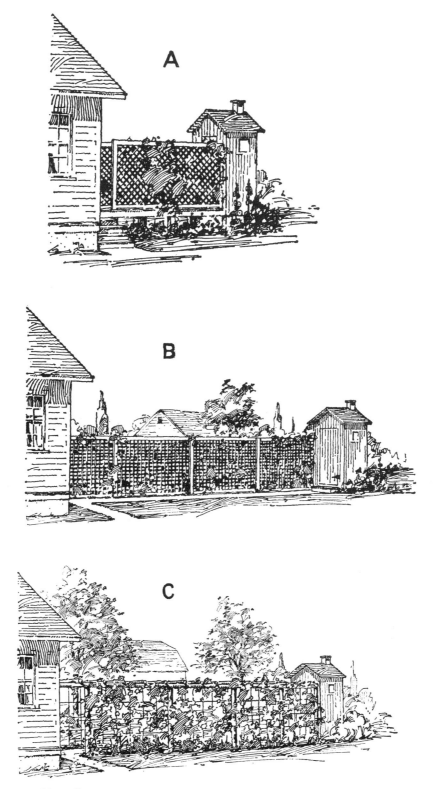

FIG. 10.—Screening the approach to a privy. *A*, Raised platform with lattice sides, suitable for short distances, convenient, and easily cleared of snow; *B*, walk hidden by latticework; *C*, walk inclosed by an arbor

of cement, 1 cubic yard of sand, and 1½ cubic yards of broken stone or screened gravel. The stone or gravel should not exceed 1 inch in diameter, except that a few cobblestones may be embedded where the vault wall is thickest, thus effecting a slight saving of materials.

A type of sanitary privy in which the excrements are received directly into a water-tight receptacle containing chemical disinfectant is meeting with considerable favor for camps, parks, rural cottages, schools, hotels, and railway stations. These chemical closets,[4] as they are called, are made in different forms and are known by various trade names. In the simplest form a sheet-metal receptacle is concealed in a small metal or wooden cabinet, and the closet is operated usually in much the same manner as the ordinary pail privy. These closets are very simple and compact, of good appearance, and easy to install or move from place to place. In another type, known as the chemical tank closet, the receptacle is a steel tank fixed in position underground or in a basement. The tank has a capacity of about 125 gallons per seat, is provided with a hand-operated agitor to secure thorough mixing of the chemical and the excretions, and the contents are bailed, pumped, or drained out from time to time.

Chemical closets, like every form of privy, should be well installed, cleanly operated, and frequently emptied, and the wastes should receive safe burial. With the exception of frequency of emptying, the same can be said

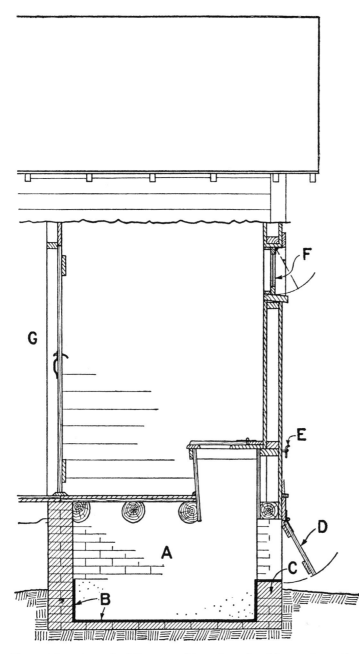

FIG. 11.—A primitive vault privy in Massachusetts. Note the tight, shallow, easily cleaned vault. *A*, Brick vault 5 by 6 feet, bottom about 1 foot in the ground; *B*, water-tight plastering; *C*, rowlock course of brick; *D*, door hinged at top; *E*, door button; *F*, three-pane window hinged at top; *G*, passageway

of chemical tank closets. With both forms of closet thorough ventilation or draft is essential, and this is obtained usually by connecting the closet vent pipe to a chimney flue or extending it well above the ridgepole of the building. The contents of the container should always be submerged and very low temperatures guarded against.

[4] Among publications on chemical closets are the following: "Chemical closets," Reprint No. 404 from the Public Health Reports, U. S. Public Health Service, June 29, 1917, pp. 1017–1020; "The chemical closet," Engineering Bulletin No. 5, Mich. State Board of Health, October, 1916; Health Bulletin, Va. Department of Health, March, 1917, pp. 214–219.

As to the germicidal results obtained in chemical closets, few data are available. A disinfecting compound may not sterilize more than a thin surface layer of the solid matter deposited. Experiments by Dr. Alvah H. Doty with various agents recommended and widely used for the bedside sterilization of feces showed "that at the end of 20 hours of exposure to the disinfectant but one-eighth of an inch of the fecal mass was disinfected." [5] Plainly, then, to destroy all bacterial and parasitic life in chemical closets three things are necessary: (1) A very powerful agent; (2) permeation of the fecal mass by the agent; (3) retention of its strength and potency until permeation is

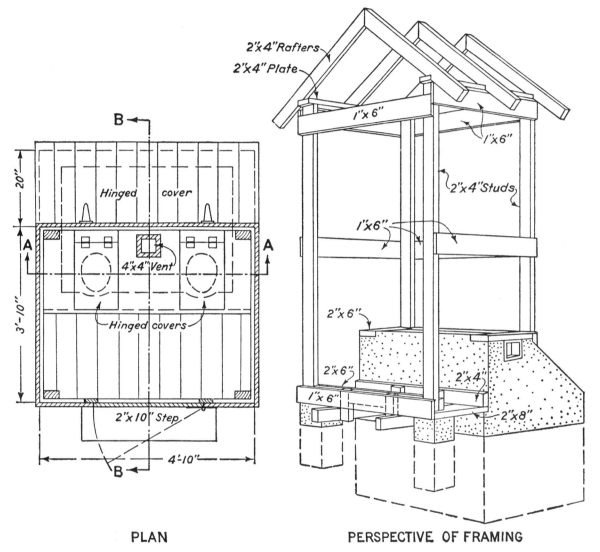

PLAN   PERSPECTIVE OF FRAMING

FIG. 12.—Two-seat vault privy

complete. The compounds or mixtures commonly used in chemical closets are of two general kinds: First, those in which some coal-tar product or other oily disinfectant is used to destroy germs and deodorize, leaving the solids little changed in form; second, those of the caustic class that dissolve the solids, which, if of sufficient strength and permeating every portion, should destroy most if not all bacterial life. Not infrequently the chemical solution is intended to accomplish disinfection, deodorization, and reduction to a liquid or semiliquid state. Ordinary caustic soda, costing about $1 in 10-pound pails, has given good results.

[5] Annual Report, Mass. State Board of Health, 1914, p. 727.

A simple type of chemical closet is shown in Figure 14, and the essential features are indicated in the notation. These closets with vent pipe and appurtenances, ready for setting up, retail for $20 and upward. A chemical tank closet, retailing for about $80 per seat, is shown in Figure 15.

The Department of Agriculture occasionally receives complaints from people who have installed chemical closets, usually on the score of odors or the cost of chemicals.

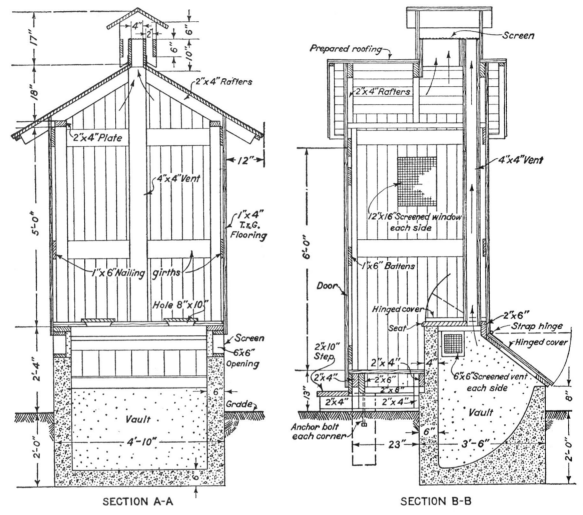

SECTION A-A   SECTION B-B

FIG. 13.—Two-seat-vault privy.   Note the shallow, water-tight, easily cleaned concrete vault

DISINFECTANTS AND DEODORANTS

Disinfection is the destruction of disease germs. Sterilization is the destruction of all germs or bacteria, both the harmful and the useful. Antisepsis is the checking or restraining of bacterial growth. Deodorization is the destruction of odor. Unfortunately in practice none of these processes may be complete. The agent may be of inferior quality, may have lost its potency, or may not reach all parts of the mass treated. A disinfectant or germicide is an agent capable of destroying disease germs; an antiseptic is an agent merely capable of arresting bacterial growth, and it may be a dilute disinfectant; a deodorant is an agent that tends to destroy odor, but whose action may consist in absorbing odor or in masking the original odor with another more agreeable one.[6]

---

[6] Those desiring more explicit information on disinfectants and the principles of disinfection are referred to U. S. Department of Agriculture Farmers' Bulletin 926, "Some Common Disinfectants," and 954, "The Disinfection of Stables," and to publications of the U. S. Public Health Service.

Of active disinfecting agents, heat from fire, live steam, and boiling water are the surest. The heat generated by the slaking of quicklime has proved effective with small quantities of excreta. Results of tests by the Massachusetts State Board of Health [7] show that the preferable method consists in adding sufficient hot water (120° to 140° F.) to cover the excrement in the receptacle,

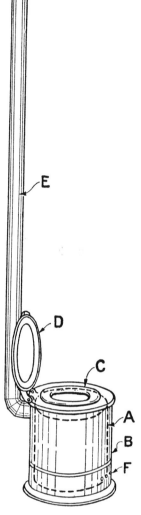

FIG. 14. — Chemical closet. *A,* Watertight sheet - metal container; *B,* metal or wooden cabinet; *C,* wooden or composition seat ring; *D,* hinged cover; *E,* 3 or 4 inch ventilating flue extending 18 inches above roof or to a chimney; *F,* air inlets

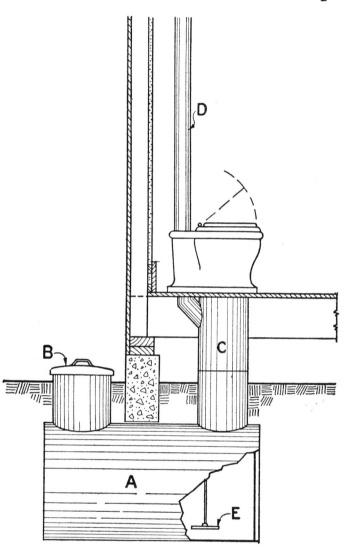

FIG. 15.—Chemical tank closet. *A,* Tank, 2 feet 3 inches by 4 feet 2 inches 5/64th-inch iron, seams welded; capacity, 125 gallons; *B,* 14-inch covered opening for recharging and emptying tank; *C,* 12-inch galvanized sheet-metal tube; *D,* 4-inch sheet-metal ventilating pipe extending above ridgepole or to a chimney; *E,* agitator or paddle

then adding small pieces of fresh strong quicklime in amount equal to about one-third of the bulk of water and excrement combined, covering the receptacle, and allowing it to stand 1½ hours or longer.

Among chemical disinfectants a strong solution of sodium hydroxide (caustic soda) or potassium hydroxide (caustic potash, lye) is very effective and is useful in dissolving grease and other organic substances. Both chemicals are costly, but caustic soda is less expensive than caustic potash and constitutes most of the ordinary commercial lyes. Chlorinated lime (chloride of lime, bleaching powder) either in solution or in powdered form is valuable. For

---

[7] Annual Report, Mass. State Board of Health, 1914, pp. 727–729.

the disinfection of stools of typhoid-fever patients the Virginia State Board of Health [8] recommends thoroughly dissolving ½ pound of best chlorinated lime in 1 gallon of water and allowing the solution to cover the feces for at least 1 hour. The solution should be kept in well-stoppered bottles and used promptly, certainly within 2 or 3 days. Copper sulphate (blue vitriol, bluestone) in a 5 per cent solution (1 pound in 2½ gallons of water) is a good but rather costly disinfectant. None of the formulas here given is to be construed as fixed and precise. Conditions may vary the proportions, as they always will vary the results. The reader should remember that few, if any, chemical disinfectants can be expected fully to disinfect or sterilize large masses of excrement unless the agent is used repeatedly and in liberal quantities or mechanical means are employed to secure thorough incorporation.

Among deodorants some of the drying powders mentioned below possess more or less disinfecting power. Chlorinated lime, though giving off an unpleasant odor of chlorine, is employed extensively. Lime in the form of either quicklime or milk of lime (whitewash) is much used and is an active disinfectant. To prepare milk of lime a small quantity of water is slowly added to good fresh quicklime in lumps. As soon as the quicklime is slaked a quantity of water, about four times the quantity of lime, is added and stirred thoroughly. When used as a whitewash the milk of lime is thinned as desired with water and kept well stirred. Liberal use of milk of lime in a vault or cesspool, though it may not disinfect the contents, is of use in checking bacterial growth and abating odor. To give the best results it should be used frequently, beginning when the vault or cesspool is empty. Iron sulphate (green vitriol, copperas) because of its affinity for ammonia and sulphides is used as a temporary deodorizer in vaults, cesspools, and drains; 1 pound dissolved in 4 gallons of water makes a solution of suitable strength.

PREVENTION OF PRIVY NUISANCE

The following is a summary of simple measures for preventing a privy from becoming a nuisance:

1. Locate the privy inconspicuously and detached from the dwelling.
2. Make the receptacle or vault small, shallow, easy of access, and water-tight.
3. Clean out the vault often. Do not wait until excrement has accumulated and decomposition is sufficiently advanced to cause strong and foul odors.
4. Sprinkle into the vault daily loose dry soil, ashes, lime, sawdust, ground gypsum (land plaster), or powdered peat or charcoal. These will absorb liquid and odor, though they may not destroy disease germs.
5. Make the privy house rain-proof; ventilate it thoroughly, and screen all openings.

OBJECTION TO PRIVIES

All the methods of waste disposal heretofore described are open to the following objections:

1. They do not take care of kitchen slops and liquid wastes incident to a pressure water system.
2. They retain filth for a considerable period of time, with probability of odors and liability of transmission of disease germs.
3. They require more personal attention and care than people generally are willing to give.

[8] Health Bulletin, Va. State Board of Health, June, 1917, pp. 277–280.

By far the most satisfactory method yet devised of caring for sewage calls for a supply of water and the flushing away of all wastes as soon as created through a water-tight sewer to a place where they undergo treatment and final disposal.

KITCHEN-SINK DRAINAGE

A necessity in every dwelling is effective disposal of the kitchen-sink slops. This necessity ordinarily arises long before the farm home is supplied with water under pressure and the conveniences that go with it. Hence the first call for information on sewage disposal is likely to relate merely to sink drainage. This waste water, though it may not be as dangerous to health as sewage containing

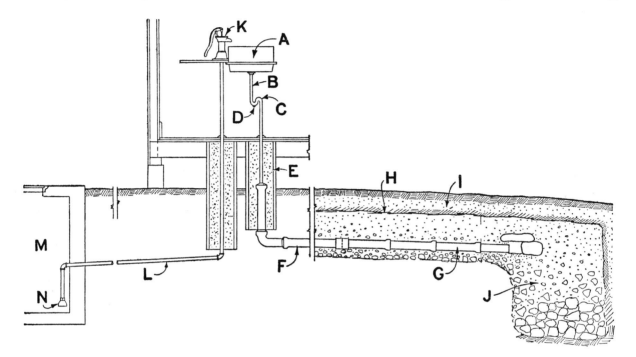

Fig. 16.—How to waste kitchen-sink drainage. *A*, Sink; *B*, waste pipe; *C*, trap; *D*, clean-out; *E*, box filled with hay, straw, sawdust, excelsior, coke, or other insulating material; *F*, 4-inch vitrified sewer pipe, hubs uphill, and joints made water-tight for at least 100 feet downhill from a well; *G*, 4-inch vitrified sewer pipe, hubs downhill, joints slightly open, laid in an 18-inch bed of coarse sand, gravel, stone, broken brick, slag, cinders, or coke; *H*, strip of tarred paper or burlap or a thin layer of hay, straw, cornstalks, brush, or sods, grass side down; *I*, 12 inches of natural soil; *J*, stone-filled pit. As here illustrated, water is drawn by a pitcher or kitchen pump (*K*) through a 1¼ or 1½ inch galvanized-iron suction pipe (*L*) from a cistern (*M*). The suction pipe should be laid below frost and on a smooth upward grade from cistern to pump and be provided with a foot valve (*N*) to keep the pump primed. If a foot valve is used, pump and pipe must be safe from frost or other means than tripping the pump be provided for draining the system

human excrements, is still a menace to the farm well and capable of creating disagreeable odor.

The usual method of disposing of sink slops is to allow them to dribble on or beneath the surface of the ground close to the house. Such drainage should be taken in a water-tight carrier at least 100 feet downhill from the well and discharged below the surface of the ground. Every sink should be provided with a suitable screen to keep all large particles out of the waste pipe. An approved form of sink strainer consists of a brass plate bolted in position over the outlet and having at least 37 perforations not larger than one-fourth inch in diameter. Provided a sink is thus equipped and is given proper care and the land has fair slope and drainage, the waste water may be conducted away through a water-tight sewer and distributed in the soil by means of a short blind drain. A simple installation, consisting of a kitchen sink and pump and means of disposal as described, is shown in Figure 16.

CESSPOOLS

Where farms have water under pressure an open or leaching cesspool is a common method of disposing of the sewage. Ordinary cesspools are circular excavations in the ground, lined with stone or brick laid without mortar. They vary from 5 to 10 feet in diameter and from 7 to 12 feet in depth. Sometimes the top is arched and capped at the ground surface by a cover of wood, stone, or cast iron. At other times the walls are carried straight up and boards or planks are laid across for a cover, and the entire structure is hidden with a hedge or shrubbery.

Except under the most favorable conditions the construction and use of a cesspool can not be condemned too strongly. They are only permissible where no other arrangement is possible. Leaching cesspools especially are open to these serious objections:

1. Unless located in porous soil, stagnation is likely to occur, and failure of the liquid to seep away may result in overflow on the surface of the ground and the creation of a nuisance and a menace.

2. They retain a mass of filth in a decomposing condition deep in the ground, where it is but slightly affected by the bacteria and air of the soil. In seeping through the ground it may be strained, but there can be no assurance that the foul liquid, with little improvement in its condition, may not pass into the ground water and pollute wells and springs situated long distances away in the direction of underground flow.

For the purpose of avoiding soil and ground-water pollution cesspools have been made of water-tight construction and the contents removed by bailing or pumping. Upon the farm, however, this type of construction has little to recommend it, for the reason that facilities for removing and disposing of the contents in a clean manner are lacking.

In some instances cesspools have been made water-tight, the outflow being effected by three or four elbows or T branches set in the masonry near the top, with the inner ends turned down below the water surface, the whole surrounded to a thickness of several feet with stone or gravel intended to act as a filtering medium. Tests of the soil water adjacent to cesspools of this type show that no reliance should be placed upon them as a means of purifying sewage, the fatal defects being constant saturation with sewage and lack of air supply. To the extent that the submerged outlets keep back grease and solid matters the scheme is of service in preventing clogging of the pores of the surrounding ground.

Where the ground about a cesspool has become clogged and waterlogged, relief is often secured by laying several lines of drain tile at shallow depth, radiating from the cesspool. The ends of the pipes within the cesspool should turn down, and it is advantageous to surround the lines of pipe with stones or coarse gravel, as shown in Figure 16 and discussed under " Septic tanks." In this way not only is the area of percolation extended, but aeration and partial purification of the sewage are effected.

Where a cesspool is located at a distance from a dwelling and there is opportunity to lead a vent pipe up the side of a shed, barn, or any stable object it is advisable to do so for purposes of ventilation. Where the conditions are less favorable it may be best, because of the odor, to omit any direct vent pipe from the cesspool and rely for ventilation on the house sewer and main soil stack extending above the roof of the house.

Cesspools should be emptied and cleaned at least once a year and the contents given safe burial or, with the requisite permission, wasted in some municipal sewerage system. After cleaning, the walls and bottom may be treated with a disinfectant or a deodorant.

SEPTIC TANKS

A tight, underground septic tank with shallow distribution of the effluent in porous soil generally is the safest and least troublesome method of treating sewage upon the farm, while at the same time more or less of the irrigating and manurial value of the sewage may be realized.

The late Professor Kinnicutt used to say that a septic tank is "simply a cesspool, regulated and controlled." The reactions described under the captions "How sewage decomposes" and "Cesspools" take place in septic tanks.

In all sewage tanks, whatever their size and shape, a portion of the solid matter, especially if the sewage contains much grease, floats as scum on the liquid, the heavier solids settle to form sludge, while finely divided solids and matter in a state of emulsion are held in suspension. If the sludge is retained in the bottom of the tank and converted or partly converted into liquids and gases, the tank is called a septic tank and the process is known as septicization. The process is sometimes spoken of as one of digestion or rotting.

History.—Prototypes of the septic tank were known in Europe nearly 50 years ago. Between 1876 and 1893 a number of closed tanks with submerged inlets and outlets embodying the principle of storage of sewage and liquefaction of the solids were built in the United States and Canada. It was later seen that many of the early claims for the septic process were extravagant. In recent years septic tanks have been used mainly in small installations, or, where employed in large installations, the form has been modified to secure digestion of the sludge in a separate compartment, thus in a measure obviating disadvantages that exist where septicization takes place in the presence of the entering fresh sewage.

Purposes.—The purposes of a septic tank are to receive all the farm sewage, as defined on page 1, hold it in a quiet state for a time, thus causing partial settlement of the solids, and by nature's processes of decomposition insure, as fully as may be, the destruction of the organic matter.

Limitations.—That a septic tank is a complete method of sewage treatment is a widespread but wrong impression. A septic tank does not eliminate odor and does not destroy all organic solids.

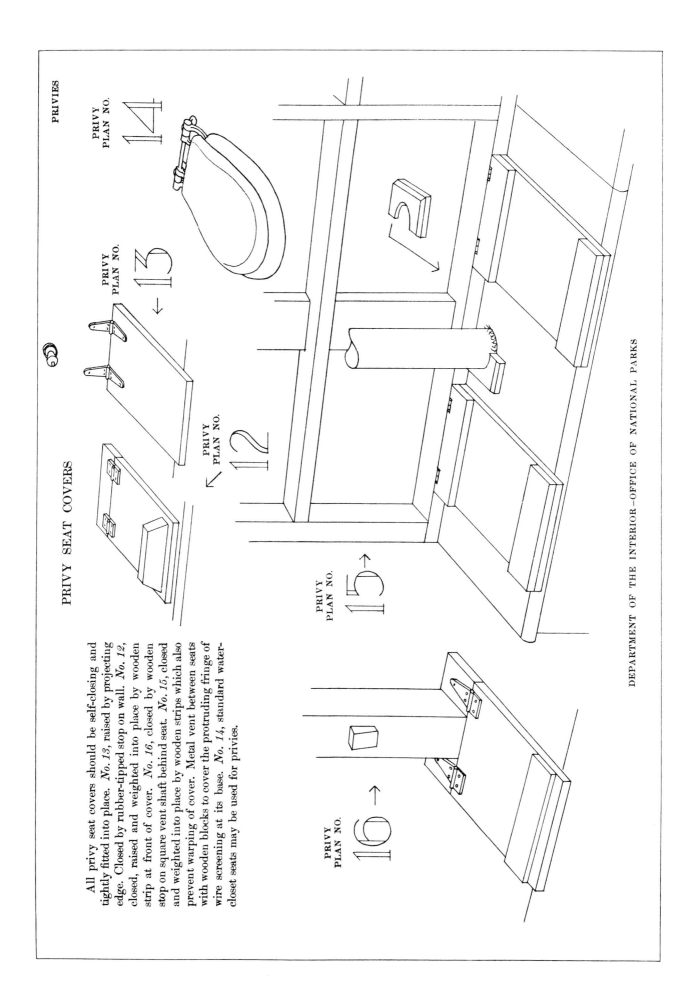

PRIVIES

PRIVY SEAT COVERS

All privy seat covers should be self-closing and tightly fitted into place. *No. 13*, raised by projecting edge. Closed by rubber-tipped stop on wall. *No. 12*, closed, raised and weighted into place by wooden strip at front of cover. *No. 16*, closed by wooden stop on square vent shaft behind seat. *No. 15*, closed and weighted into place by wooden strips which also prevent warping of cover. Metal vent between seats with wooden blocks to cover the protruding fringe of wire screening at its base. *No. 14*, standard water-closet seats may be used for privies.

PRIVY PLAN NO. 14

PRIVY PLAN NO. 13

PRIVY PLAN NO. 12

PRIVY PLAN NO. 15

PRIVY PLAN NO. 16

DEPARTMENT OF THE INTERIOR—OFFICE OF NATIONAL PARKS

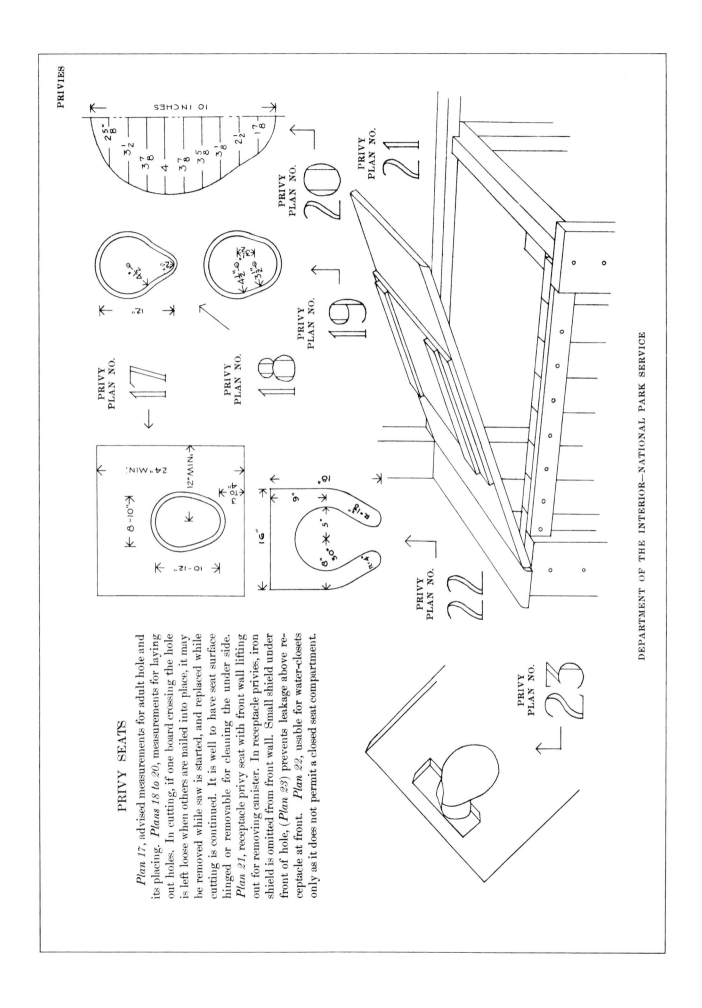

PRIVIES

10 INCHES

PRIVY PLAN NO. 21

PRIVY PLAN NO. 20

PRIVY PLAN NO. 19

PRIVY PLAN NO. 18

PRIVY PLAN NO. 17

12"

24" MIN.

12" MIN.

8-10"

10-12"

16"

9"

10"

PRIVY PLAN NO. 22

PRIVY PLAN NO. 23

DEPARTMENT OF THE INTERIOR—NATIONAL PARK SERVICE

PRIVY SEATS

Plan 17, advised measurements for adult hole and its placing. Plans 18 to 20, measurements for laying out holes. In cutting, if one board crossing the hole is left loose when others are nailed into place, it may be removed while saw is started, and replaced while cutting is continued. It is well to have seat surface hinged or removable for cleaning the under side. Plan 21, receptacle privy seat with front wall lifting out for removing canister. In receptacle privies, iron shield is omitted from front wall. Small shield under front of hole, (Plan 23) prevents leakage above receptacle at front. Plan 22, usable for water-closets only as it does not permit a closed seat compartment.

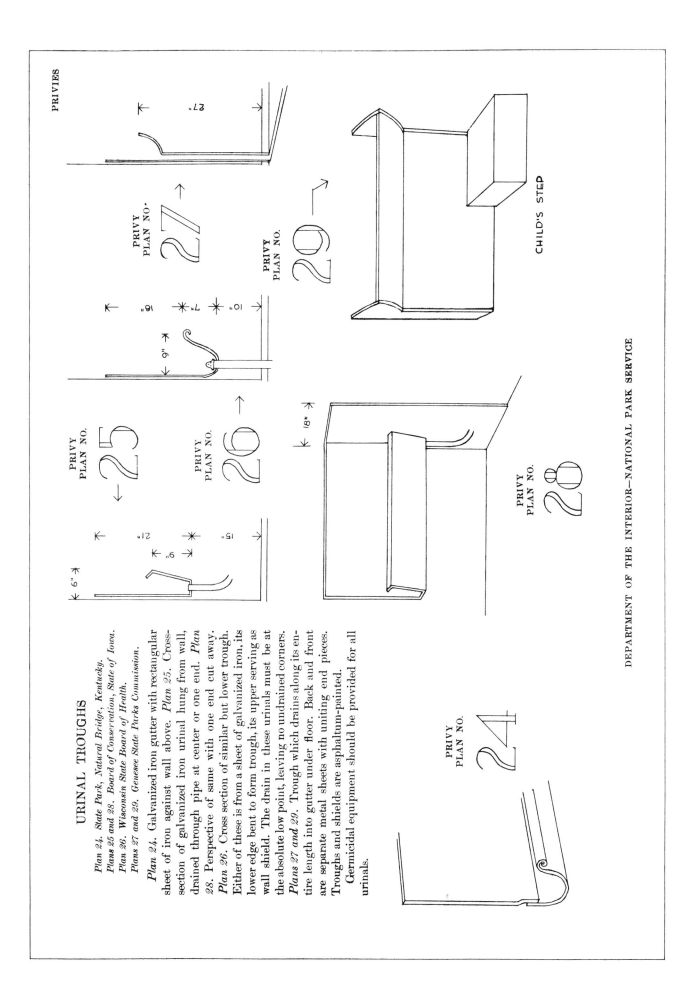

URINAL TROUGHS

Plan 24. State Park, Natural Bridge, Kentucky.
Plans 25 and 28. Board of Conservation, State of Iowa.
Plan 26. Wisconsin State Board of Health.
Plans 27 and 29. Genesee State Parks Commission.

Plan 24. Galvanized iron gutter with rectangular sheet of iron against wall above. *Plan 25.* Cross-section of galvanized iron urinal hung from wall, drained through pipe at center or one end. *Plan 28.* Perspective of same with one end cut away. *Plan 26.* Cross section of similar but lower trough. Either of these is from a sheet of galvanized iron, its lower edge bent to form trough, its upper serving as wall shield. The drain in these urinals must be at the absolute low point, leaving no undrained corners. *Plans 27 and 29.* Trough which drains along its entire length into gutter under floor. Back and front are separate metal sheets with uniting end pieces. Troughs and shields are asphaltum-painted.

Germicidal equipment should be provided for all urinals.

CHILD'S STEP

PRIVY PLAN NO. 27

PRIVY PLAN NO. 29

PRIVY PLAN NO. 25

PRIVY PLAN NO. 26

PRIVY PLAN NO. 28

PRIVY PLAN NO. 24

DEPARTMENT OF THE INTERIOR—NATIONAL PARK SERVICE

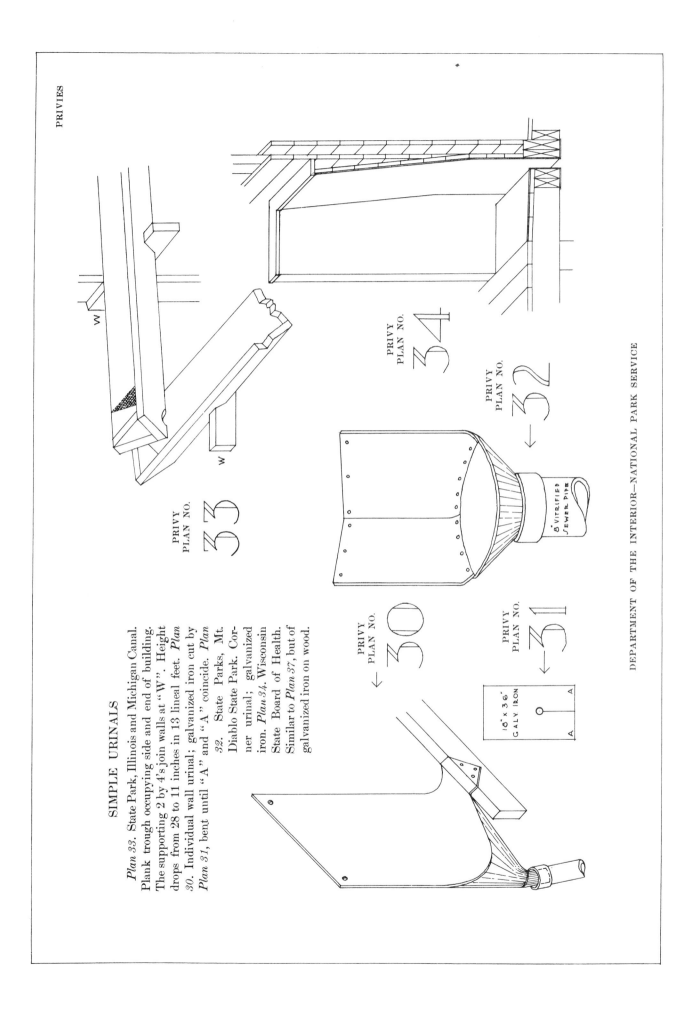

SIMPLE URINALS

Plan 33. State Park, Illinois and Michigan Canal. Plank trough occupying side and end of building. The supporting 2 by 4's join walls at "W". Height drops from 28 to 11 inches in 13 lineal feet. *Plan 30.* Individual wall urinal; galvanized iron cut by *Plan 31,* bent until "A" and "A" coincide. *Plan 32.* State Parks, Mt. Diablo State Park. Corner urinal; galvanized iron. *Plan 34.* Wisconsin State Board of Health. Similar to *Plan 37,* but of galvanized iron on wood.

PRIVY
PLAN NO.
33

PRIVY
PLAN NO.
34

PRIVY
PLAN NO.
32

PRIVY
PLAN NO.
30

PRIVY
PLAN NO.
31

8" VITRIFIED SEWER PIPE

10" x 36"
GALV. IRON

DEPARTMENT OF THE INTERIOR—NATIONAL PARK SERVICE

CONCRETE WALL URINALS AND AUXILIARY CAMP URINAL

Plan 35. New York State Board of Health.
Plans 36 and 37. Wisconsin State Board of Health.

Plan 35. Auxiliary urinal for night use near tents. Lengths of 2-inch galvanized pipe extend down into dry well which is earth-covered to prevent rising odors. In top of each pipe is a screened funnel. Whitened stones mark the area for nocturnal visibility. *Plans 36 and 37.* Concrete wall urinals. *Plan 37.* Sloping wall; spattering less objectionable than vertical one. Floorless gutter opening directly into pit avoids the clogging to which *Plan 36* is susceptible, but makes it difficult to exclude flies from the pit and to eliminate pit odors from room. The Wisconsin Board prefer the closed gutter (*Plan 36*) sprinkled profusely with lime, its floor sloping to trap drains which have ceiling vent directly above.

Exposed urinal walls are finished with a rich cement coating water-proofed with asphaltum. In *Plan 36* urinal wall serves also as partition, toilet seats backing against opposite side. One pit serves seats and urinal.

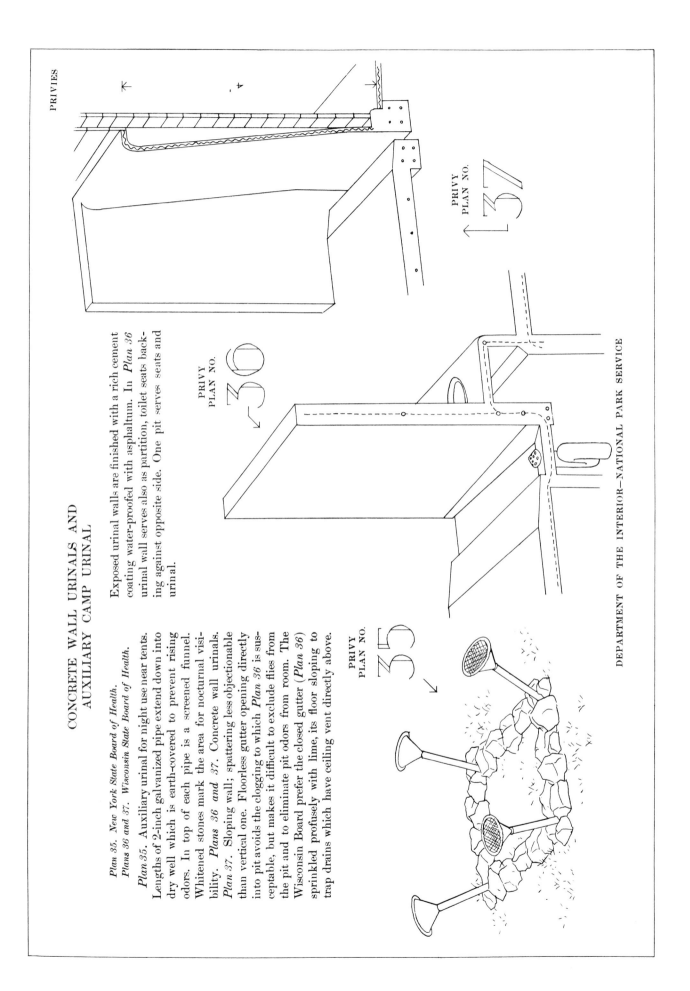

PRIVY PLAN NO. 37

PRIVY PLAN NO. 36

PRIVY PLAN NO. 35

DEPARTMENT OF THE INTERIOR–NATIONAL PARK SERVICE

131

VENTILATOR HOODS.

The top of a vent should be hooded to exclude light and rain, and to prevent a reversal of air current. *Plans 38, 39, and 40.* Metal vent pipes from toilet pit. *Plans 41, 42, and 43.* Square wooden shafts from pit. *Plan 44.* Square wooden shaft from pit and an adjacent oblong shaft which extends through roof of building only, airing interior of room. *Plan 38.* Formed from end of pipe itself by cutting and bending as shown. Metal vent pipes are commonly screened at base. Cut hole the size of pipe in seat surface, cover with sheet of wire mesh and force pipe down into place. Wooden vents are screened at either top or bottom. *Plan 44.* Hood 6 inches wider than shaft. When in place with ends resting on roof, its horizontal connecting boards are level at the top with the shaft, but are 3 inches from it.

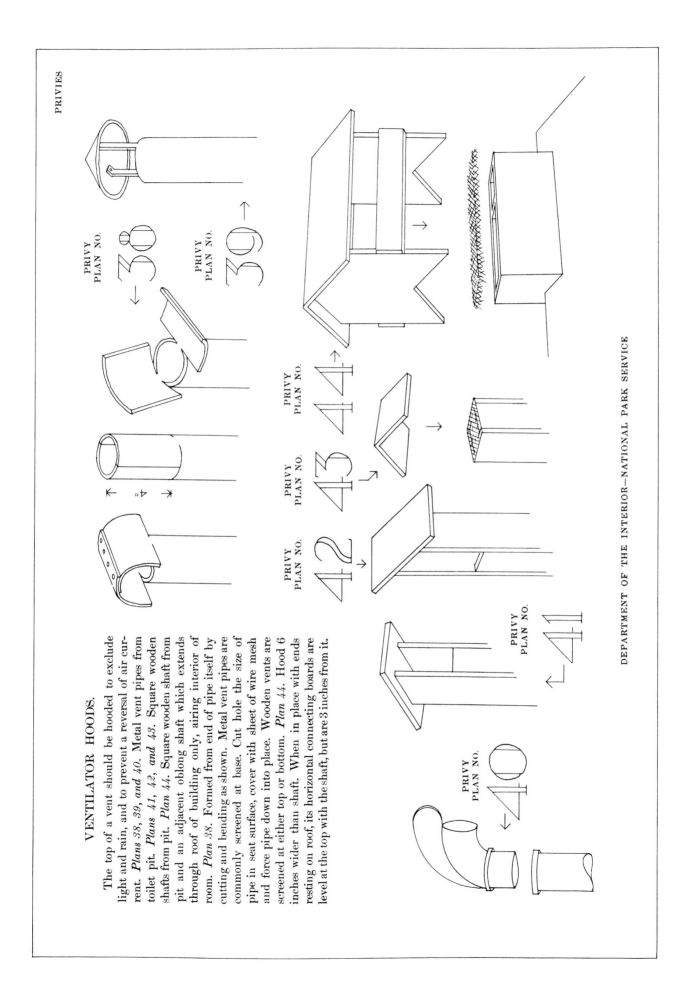

PRIVY PLAN NO. 38

PRIVY PLAN NO. 39

PRIVY PLAN NO. 40

PRIVY PLAN NO. 41

PRIVY PLAN NO. 42

PRIVY PLAN NO. 43

PRIVY PLAN NO. 44

DEPARTMENT OF THE INTERIOR—NATIONAL PARK SERVICE

MISCELLANEOUS PRIVY DETAILS

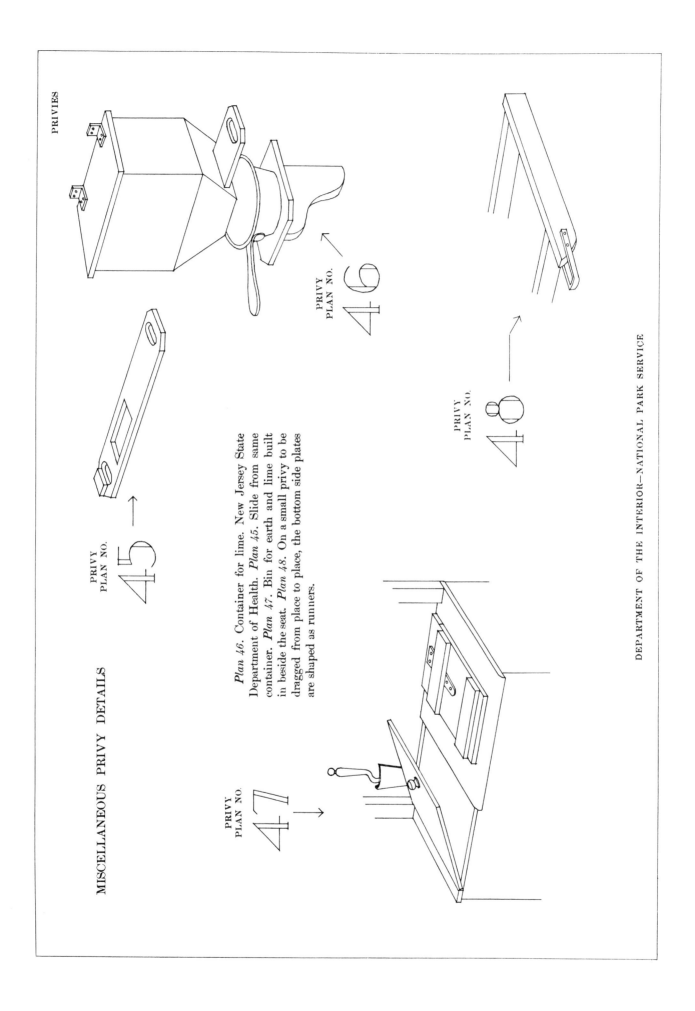

PRIVY
PLAN NO.
45

PRIVY
PLAN NO.
46

PRIVY
PLAN NO.
47

PRIVY
PLAN NO.
48

Plan 46. Container for lime. New Jersey State Department of Health. *Plan 45.* Slide from same container. *Plan 47.* Bin for earth and lime built in beside the seat. *Plan 48.* On a small privy to be dragged from place to place, the bottom side plates are shaped as runners.

DEPARTMENT OF THE INTERIOR—NATIONAL PARK SERVICE

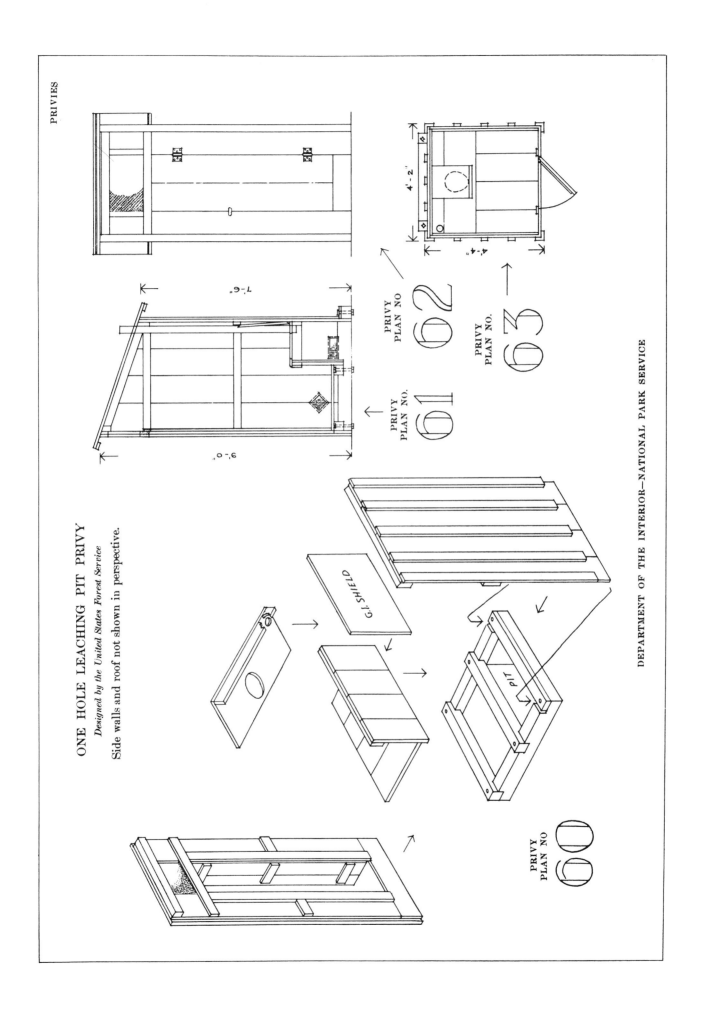

ONE HOLE LEACHING PIT PRIVY

Designed by the United States Forest Service

Side walls and roof not shown in perspective.

PRIVY PLAN NO. 60

PRIVY PLAN NO. 61

PRIVY PLAN NO. 62

PRIVY PLAN NO. 63

G.I. SHIELD

PIT

7'-6"

9'-0"

4'-2"

4'-4"

DEPARTMENT OF THE INTERIOR—NATIONAL PARK SERVICE

134

PRIVY WITH EXTERIOR ENTRANCE FOR EACH COMPARTMENT

Designed by Akron Metropolitan Park Board.

In this privy each compartment is a separate room with exterior entrance. This makes it possible to eliminate passage or waiting room and to construct a consequently narrower building. Partitions between rooms do not extend to floor or ceiling. Each room is ventilated by screened louvres over the door and screened holes below the rear eaves. No pit ventilation is attempted on the grounds that, in the opinion of H. S. Wagner, director-secretary, it is an added expense and seldom satisfactory; that it must be dependent on closed seat covers and they are never certain. This pit is, therefore, solidly constructed with no screening to be kept in repair. An unventilated pit requires scrupulous maintenance. The Akron Park Board spray floor and vault interior frequently with creosote. They use lime in leaching pits.

Commercial wooden toilet seats of the type used for water closets are superimposed over rectangular openings in the seat surface. The exterior of the building is of chestnut bark shingles laid over waterproof building paper. These give an attractive yet inconspicuous character in the wooded locations where the buildings are commonly placed. Each door has three bark panels, and an upper section of translucent wire-glass to admit light.

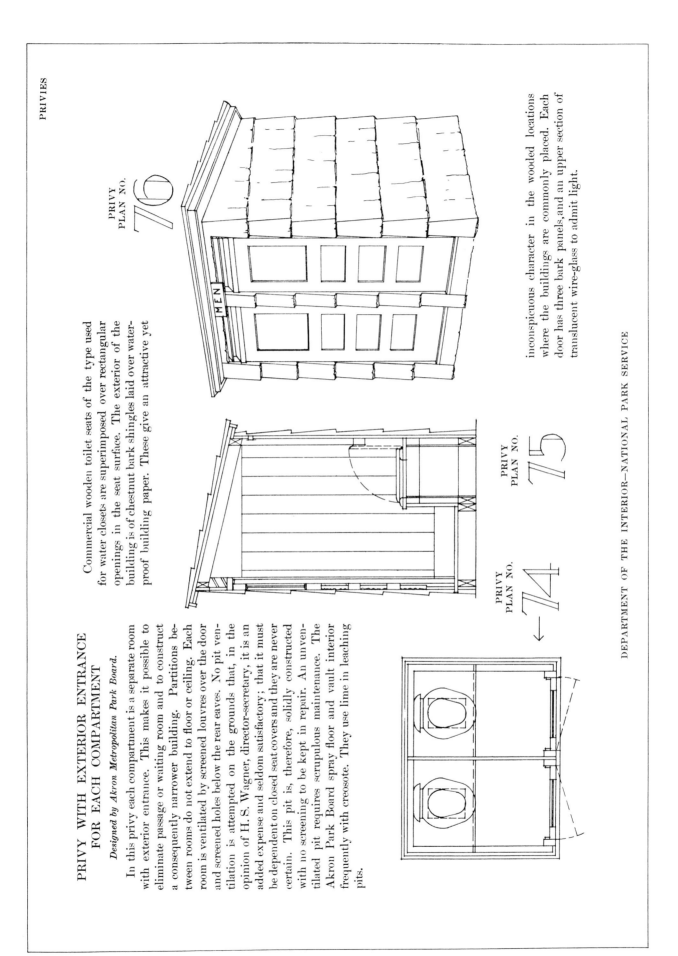

PRIVY PLAN NO. 76

PRIVY PLAN NO. 75

PRIVY PLAN NO. 74

DEPARTMENT OF THE INTERIOR—NATIONAL PARK SERVICE

135

The Old Backhouse

When memory keeps me company
  and moves to smiles or tears,
A weather-beaten object looms
  through the mist of years.
Behind the house and barn it stood,
  a hundred yards or more,
And hurrying feet a path had made,
  straight to its swinging door.

Its architecture was a form
  of simple classic art,
But in the tragedy of life
  it played a leading part.
And oft the passing traveler
  drove slow, and heaved a sigh,
To see the modest hired girl
  slip out with glances shy.

We had our posey garden
  that the women loved so well,
I loved it, too, but better still
  I loved the stronger smell
That filled the evening breezes
  so full of homely cheer,
And told the night-o'ertaken tramp
  that human life was near.

On lazy August afternoons,
  it made a little bower,
Delightful, where my grandsire sat
  and whiled away an hour.
For there the morning-glory
  its very eaves entwined,
And berry bushes reddened
  in the steaming soil behind.

All day fat spiders spun their webs
  to catch the buzzing flies
That flitted to and from the house,
  where Ma was baking pies.
And once a swarm of hornets bold,
  had built a palace there,
And stung my unsuspecting aunt —
  I must not tell you where.

Then Father took a flaming pole —
  that was a happy day —
He nearly burned the building up,
  but the hornets left to stay.
When summer blooms began to fade
  and winter to carouse
We banked the little building with
  a heap of hemlock boughs.

But when the crust was on the snow
  and the sullen skies were gray,
In sooth the building was no place
  where one would wish to stay.
We did our duties promptly,
  there one purpose swayed the mind;
We tarried not, nor lingered long
  on what we left behind.

The torture of that icy seat
  would make a Spartan sob,
For needs must scrape the goose flesh
  with a lacerating cob,
That from a frost-encrusted nail,
  was suspended by a string —
My Father was a frugal man
  and wasted not a thing.

When grandpa had to "go out back"
  and make his morning call,
We'd bundle up the dear old man
  with a muffler and a shawl,
I knew the hole on which he sat —
  'twas padded all around,
And once I dared to sit there —
  'twas all too wide I found.

My loins were all too little,
  and I jackknifed there to stay,
They had to come and cut me out,
  or I'd have passed away.
Then Father said ambition
  was a thing that boys should shun,
And I must use the children's hole
  'till childhood days were done.

And still I marvel at the craft
  that cut those holes so true.
The baby hole, and the slender hole
  that fitted Sister Sue.
That dear old country landmark;
  I've tramped around a bit,
And in the lap of luxury
  my lot has been to sit,

But ere I die I'll eat the fruit
  of trees I've robbed of yore,
Then seek the shanty where my name
  is carved upon the door.
I ween the old familiar smell
  will sooth my jaded soul,
I'm now a man but none the less,
  I'll try the children's hole.

— James Whitcomb Riley

The Vanishing American BARBER SHOP

In the most comprehensive book ever written on barber shop history, Ronald S. Barlow traces the evolution of professional shaving and haircutting from ancient Egypt, Greece, and Rome to the uni-sex shops of the 1960's. Scores of vintage photographs and hundreds of illustrations accompany a text which is both educational and entertaining. Few people are aware that before World War I most men did not shave themselves. They visited their local barber shop for whisker removal at least twice a week. The shaving parlor was a men's club of sorts, a bay rum-scented repository of spicy maga-zines and the latest sporting news where butchers, bakers, bankers, carpenters, and cattlemen rubbed elbows and puffed on five-cent cigars while waiting for a turn in the red leather chair. Several chapters in this book are devoted to the memoirs of old-time barbers who actually worked in shops from 1830 to 1950. Also included are lessons on shaving (with a straight razor) and haircutting as taught in early barber school manuals. Doctors and dentists will discover the origins of their profes-sions in a section on bloodletting monks and barber-surgeons of the 15th century. One hundred bea-utifully engraved 9 x 12-inch pages are faithfully reproduced from rare barber-supply catalogs of the 1880–1920 period. Antique dealers, museum curators and "barberiana" collectors will find the price guide section an invaluable research tool. Included is an extensive listing of old-time occupa-tional shaving mugs, which are currently selling for up to $5,000 each. **224 pages. 9 x 12 paperback.** **ISBN 933846-04-5. $16.95 plus $2 postage.**

The Vanishing American Outhouse

Rave reviews in *The New York Times, Los Angeles Times, and Washington Post.* "A funny, fascinating book," says the American Library Association. For some unexplained reason, scholars have ignored (or avoided) this important part of our architectural heritage, and only recently have historians begun to appreciate the "necessary house" as a true form of folk architecture. Privies which were once routinely burned or demolished are now quickly sold to landscape designers who move these quaint buildings into the backyards of wealthy clients. The book contains nearly 200 col-or photographs and plan drawings of American privies constructed between 1820 and 1940, plus a collection of rarely seen privy postcards. Also included is a reprint section containing U.S. Govern-ment pamphlets on outhouse construction, and a bibliography which features very candid reviews of nearly every outhouse book ever published. The 9 x 12 inch format and extra large typeface make this colorful paperback an excellent gift for the over-fifty crowd, and an enlightening cultural history for baby boomers who have never used an authentic outdoor convenience. **144 pages. 9 x 12 paperback. ISBN 933846-02-9. $15.95 plus $2 postage.**

The Antique Tool Collector's Guide to Value

Old woodworking tools are one of today's hottest new areas of collecting. Long neglected by all but a few sophisticated insiders, this field is growing rapidly. Several Stanley/Bailey carpenter planes are selling in the $500–$900 range, and an early plow plane sold for over $6,000 at a recent tool auction. Ronald Barlow spent three years working on this guide . . . recording dealer and auction prices from all over the world for every kind of old tool imaginable. Just a few of the over-5,000 items described and priced: adzes, anvils, axes, bitstocks, chisels, cobbler's tools, cooper's items, hammers, hat-chets, levels, planes, plumb bobs, rules, saws, screwdrivers, spoke shaves, surveying instruments, tinsmith's tools, wrenches & wheelwright's tools. Contains historical data on inventors and manufacturers. Bibliography includes names and addresses of antique tool auctioneers, dealers, and collector organizations. Over 2,000 illustrations. Extra-large 9 x 12 inch paperback volume. A best seller, over 75,000 copies now in print. (Includes 1991 Auction Update.) **236 pages. 9 x 12 paperback. ISBN 933846-01-0. $12.95 plus $2 postage.**

How To Be Successful In The Antique Business

OPENING AN ANTIQUE SHOP? This 185-page Start-up Manual can help you to succeed where others often fail. Some of the dozens of chapters are: Wholesale Antique Outlets, Nationwide List of Wholesale Giftware Centers, How to Purchase Estates, How to Become a Licensed Appraiser, How to Make a Living at Antique Shows, An Introduction to the Jewelry Business, Professional Furniture Refinishing & Restoring, Consignment Sales, Cash Flow & Profit Projections, Shopping Center Leases, Mail-Order Selling, Proven Advertising Techniques, Container Importing from Europe, Dealer Trade Publications, Selling Rare Books & Prints, Finding and Dealing with "Pickers," plus over 45 charts, photos, and sample forms. (Fifth printing.) **185 pages. 6 x 9 paperback. ISBN 933846-00-2. $9.95 plus $2 postage.**

Victorian Houseware, Hardware and Kitchenware

Factory-made household items of the 1875 to 1905 period are now avidly sought after by serious collectors. Ronald S. Barlow (author of *The Antique Tool Collector's Guide to Value*) and Linda Franklin (author of *300 Years of Kitchen Collectibles*) have combined talents to produce a giant 376-page price guide containing over 2,000 illustrations. This new book will help you appraise such items as: Andirons, Apple Corers, Bells, Brassware, Boxes, Butter Prints, Cake Baskets, Can Openers, Cast Iron Cookware, Cork Screws, Clocks, Coffee Mills, Door Knobs, Hardware, Egg Beaters, Graniteware, Kitchen Tools, Locks and Keys, Silverware, Tableware, Tea Kettles, Oil Lamps, Lanterns, Match Safes, Mouse Traps, Napkin Rings, Sad Irons, Iceboxes, Salt Spoons, Scis-sors, Sewing Machines, Thimbles, Trivets, Tea Pots, Tinware, Tobacco Cutters, Washing Machines, Woodstoves, Wicker Furniture and Farm Tools. All this information, plus copyright-free use of the wonderful old wood engravings makes this book a valuable tool for any artist, craftsman, antique dealer, or collector. **376 pages. 9 x 12 paperback. ISBN 933846-03-7. $19.95 plus $2 postage.**

Order from your favorite bookstore, or directly from Windmill Publishing Co., 2147 Windmill View Road, El Cajon, California 92020